The

Specter Matrix

by

Thomas Velsun

This novel is a work of fiction. Names, characters, places, events, and incidents are either products of the author's imagination or are used fictitiously. Any resemblance to actual persons, living or dead, or actual events is purely coincidental. Please don't copy, share, or use any part of this book without written permission from the author. Short quotes for reviews are okay if credit is given.

About the Author:

Thomas Velsun lives with his wife in coastal Florida. He's a former Army officer and a veteran of the Vietnam War. After leaving the Military he spent forty years in business and retired from a large global technology company.

The knowledge gained from his three college degrees from three different major universities (a BS in Math and Accounting, an MS in Engineering, and an MBA) has contributed heavily to his success in business and to his grasp of the subject matter in his novels.

During his business career, he completed special projects with various government agencies in Washington DC, including the FBI,CIA,US Mint, and IRS. He has participated in special projects with RAND Corporation (a think tank formed after World War II to continue the activities of various experts accumulated during the war years). He has worked in many different areas in the United States and in foreign countries. He has researched many conspiracy theories and has included selected subject matter from some of them in his novels.

He's a former athlete and maintains fitness with daily workouts involving weightlifting. He also enjoys playing golf and tennis.

The Specter Matrix

Chapter 1

In Jacksonville, Florida, Matt Baker, Vice President of Tuxtun Industries knew he on a target list compiled by the KGB and he constantly stayed on high alert.

Matt and his boss, Ralph Gibson, President of Tuxtun Industries, were kept well informed by the CIA and they both were in the habit of constantly staying extra alert for trouble since they had already survived many previous skirmishes with assassins.

Right now, Matt had had completed his shopping and was walking toward his car in the parking lot.

Six feet tall, agile, and a muscular-205 pounds, Matt felt he could cope with most any villain in physical combat but getting hit with a sniper's bullet was a different problem. He scanned the parking area again and picked up his pace.

His meetings and training sessions at the local FBI facilities and the CIA facilities in McLean, Virginia, were helpful. He was well practiced on many methods of defense.

Matt stayed on full alert for danger as he continued toward his car. Within moments, a large figure emerged from behind a parked SUV at the end of the row of cars beside him.

The stranger walked slowly forward.

Matt watched the stranger halt about five feet away.

The stranger didn't say anything as he stared at him.

Matt immediately concluded the man was a professional, but so was he. He had acquired a cool demeanor while surviving many tough assignments working on several large projects with Law Enforcement. He had also developed superior tactical skills from rigorous training with Law Enforcement.

The stranger wore a tan sports coat over a blue knit shirt. His casual slacks and shoes were dark brown. Thick muscles made his clothes a tight fit. A slight bulge in the stranger's coat protruded under his left arm.

So, he's armed, Matt thought, but so am I.

"Is something on your mind?" Matt asked.

"So, you're the famous Matt Baker." The stranger grinned. "You're probably not as tough as I've heard you are."

"Who're you getting your information from?"

The stranger shrugged. "Some people I work for."

"How did you know I would be here?"

"My bosses have a lot of advanced technology. We can track you anywhere you go."

Matt narrowed his eyes. "You know my name, what's yours?"

The stranger made no comment. His grin expanded.

There was a brief silence.

"What do you want?" Matt asked.

"You're a major nuisance to my employers and I've been instructed to give you one last warning to back off some things you're going to be doing."

"Such as?"

"You should already know." The stranger's voice was business-like, cold. "If you don't know right now, you'll find out soon."

Under his navy-blue sports coat, Matt's 9mm Glock 19 rested in a black nylon holster toward the back of his right hip. His well-honed skill of being able to draw and fire within half a second was useful when needed. Keeping his balance on both feet, he stayed ready for trouble.

The stranger remained silent and his grin remained intact.

Keeping most of his attention on the stranger, Matt again scanned the premises around him. He knew he was a sitting duck for a sniper, but he detected no other threats now.

"Do you have any more to say?" Matt asked.

"I'm good at my job and I've handled all my previous assignments well."

"Anything else?"

"Maybe," the stranger said. "I'll get around to it. You need to exercise a little patience."

Matt considered all his options. Exercising a little patience was indeed what he was doing, but that had its limits. "If you have any more comments, I'm listening."

"Sticking your nose into certain things disturbs my bosses."

"Who are your bosses?"

"I'm not going to tell you that." The stranger pursed his lips and looked down. It appeared as if he were thinking about something.

"My bosses want me to emphasize this is your final warning," the stranger said.

"And you're not going to give me any specifics?"

"Like I keep saying, you should already know." The stranger narrowed his eyes. "How clear do I need to make it? Are you stupid or something?"

Matt didn't comment. The conversation had reached a ridiculous stage.

Staying ready for anything, Matt continued scanning the area while keeping part of his gaze on the stranger.

"The message has been delivered," the stranger said as he turned and started walking toward another area of the parking lot.

"There could be some argument about how clear your message has been."

"You want to file a complaint?" The stranger stopped and

turned back toward him. A slight grin tugged at the corners of his mouth as he widened his stance. He narrowed his eyes and focused directly on Matt. His grin expanded.

Matt stayed ready for action. He wasn't sure of the stranger's true intentions, and he didn't want to be caught off guard.

Matt felt his phone vibrate in his pocket. "I'm getting a phone call," he said, "probably my boss."

The stranger shrugged. "Go ahead and take it. You can tell your boss you're talking to me if you want."

While Matt reached for his phone with his left hand, he filled his right hand with his Glock 19 in less than a second. He extended his pistol. "Hands in the air or you'll die on the spot."

The stranger exhaled sharply and slowly raised his arms.

Matt lifted his phone to his left ear. "Baker."

"Where are you?" Ralph asked.

"In the parking lot at the hardware store where we both shop, just leaving," Matt said, keeping his pistol steady and pointed at the stranger. "I have company."

"Oh."

Matt explained.

"Justin is here and I'm sure he'll send someone right over to take care of the situation," Ralph said.

Matt heard Ralph disconnect.

The stranger kept his hands in the air and didn't seem worried.

*****

An hour later, Matt arrived at Ralph's office where everyone was gathered at the round conference table with seating for eight.

Ralph gestured for Matt to join them.

Matt walked to the table and joined Ralph, Judy Caldwell, Tuxtun Business Analyst, Todd Green, Tuxtun technology specialist, Steve Baxter, Tuxtun Security Chief, Justin Mason,

Jacksonville Police Chief, Ben Fulton, FBI Special Agent in Charge (SAC) of the Jacksonville field office, and Kyle Rankin, the top technology specialist in the FBI field office.

Ben gestured toward Kyle. "He just finished a series of meetings with various specialists at RAND. There are some new issues that I thought you should know about and I wanted to have this meeting to keep all of you fully informed."

"Do any of them relate to directly to Tuxtun?" Ralph asked.

Ben gave a slow nod. "Some could directly relate to all of us very soon and in a big way."

Ralph gave a slight shrug. "Go ahead and give us some details."

"I'll summarize and then any of you can ask questions when you want more specifics," Ben said.

Matt and Ralph nodded.

Ben narrowed his eyes. "There is something called the Specter Matrix that has come to our attention. Our intelligence agencies tell us the Hidden Empire uses that term when they refer to the way their money flows through the global financial systems."

"Specter as in phantom or ghost?" Matt asked.

"Yep," Ben said as he gave a short laugh. "They obviously want all the various payments they make to institutions and individuals to be well hidden. He motioned toward Kyle.

Kyle leaned forward. "While discussing the situation with RAND I found out that the Specter Matrix primarily involves elements of the World Bank, the IMF, the WTO, and the Federal Reserve."

A brief silence ensued.

"The IMF is the International Monetary Fund, and the WTO is the World Trade Organization, right?" Steve asked.

"Right," Kyle said. "You have a good memory about some of our previous discussions." He paused for a moment. "And remember the World Bank has five different divisions. I'll explain more about how they fit into this later."

Steve nodded and didn't comment.

"Another interesting thing is that the Specter Matrix includes multiple accounts established on the Isle of Man and the Channel Islands, but those accounts are never used," Kyle said. "I'll explain more about that later. It gets very complicated quickly. In short, various aspects of all the global monetary organizations and various aspects of all the money laundering havens are used by the Hidden Empire to secretly funnel money to activities and organizations they want to support." Kyle narrowed his eyes. "And it looks like they also use it to receive money, and they use a different path for the money flow every time."

"How did you discover the Hidden Empire used the name Specter Matrix in reference to their money flow?" Matt asked.

"NSA has improved their technology. They now have access to several quantum computers. All phone calls can now be intercepted and decrypted no matter how good the security is or how good the encryption is," Kyle said. "We intercepted a lot of phone calls that referenced the Specter Matrix, and we eventually put a lot of the pieces of the puzzle together."

A brief silence ensued.

"Tell me again: exactly where are the Channel Islands?" Steve asked.

"The Channel Islands are the islands of Jersey and Guernsey off the coast of England, located in the English Channel," Kyle said. "There are some smaller islands in the mix, but they're considered a part of the bigger islands. They're all closer to France than Great Britain. They're thirteen miles from the northern coast of France and 85 miles from the southern coast of Great Britain." Kyle scanned the faces around the table. "These islands have provided a variety of tax havens, immune to inquiries from foreign authorities, but not immune to the CIA."

"And again, where's the Isle of Man?" Steve asked.

"North of the Channel Islands, it has a central location in the Irish Sea between Northern Ireland and Great Britain," Kyle said. "The financial community operates, for the most part, the same as it does in the Channel Islands."

Ralph raised his right hand. "Okay, I think we have enough details on the broader financial situation. Now let's concentrate on identifying what we think will directly affect Tuxtun."

"There's going to be a meeting of some billionaires, each from a different country, on Jekyll Island. In our investigation, we found that Leon Schulz is still alive. He along with Jake Bolton and some leaders of large drug cartels are also going to attend that meeting," Ben said. "The discussion will reference goals and methods that were important to Hugo Wagner and to Leon, both being descendants of Hitler and previously trying to conquer the world."

There was a brief silence.

"Since Leon Schulz survived his tenure at Belzex, he's probably going to lead the discussion in that meeting." Ben stared at Matt and Ralph for several moments. "We have to assume he's going to be back to his old tricks."

Another brief silence ensued.

Ralph glanced at Ben and narrowed his eyes. "So how do you think this will affect Tuxtun?"

"We found out the billionaires have banded together, and all are working with some large drug cartels since that provides them one of their best sources of big money," Kyle said. "You'll probably encounter the cartels in some of your dealings in growing poppies, and there will likely be a conflict. If there's a conflict, the cartels will want to get you out of their way, and they employ a lot of assassins."

Matt and Ralph both nodded but neither seemed convinced.

Ben looked at Ralph. "You're currently expanding your business for growing poppies and producing legal products like Morphine and Codeine for pharmaceutical uses, right?"

"Right," Ralph said. "We're in the process of creating a big market with pharmaceutical companies. We've been substantially increasing our profits lately."

"I see a lot of big conflicts that might happen," Ben said.

He leaned forward and explained.

Matt and Ralph each tightened his lips and made no immediate comment.

"Those two drugs you manufacture are made from poppy straw, aren't they?" Kyle asked.

"Concentrate of poppy straw," Matt said. "It's referred to as CPS and it's every bit legal as long as it's for producing legal drugs."

"You'll be competing with the drug cartels on growing the poppies and producing the CPS." Ben gave a slight shrug. "They're producing heroin and opium for big profits, and they'll likely consider you to be in their way for making even bigger profits. All of them might get irritated about some things you'll do and try to knock you off."

"I'm guessing the Hidden Empire is still active in world affairs?" Matt asked. "Do you know if they figure into this situation in any way?"

Ben gestured to Kyle.

"They probably do, but we aren't sure about the details yet." Kyle took a deep breath and exhaled slowly. "It's difficult to collect any information about them. We know they must have a lot of money coming in to maintain their influence. If the source of some of that money is from illegal drug operations you might end up causing them a problem and they won't sit still and allow you to interfere with any of their incoming money supply."

Ralph glanced at Ben and leaned forward. "So where does that leave us?"

"Very likely with a big problem," Ben said, "if you continue your business the way I think you plan to. I wanted to ensure you were forewarned."

Matt glanced at Ben and Kyle. "We didn't seem to get in the way of the Hidden Empire in our dealings earlier."

"Since Leon Schulz is still around, that's not a closed book yet," Kyle said.

Ben nodded.

"And, at that time, you weren't very far along into a big expansion of your legal business for producing more Codeine and Morphine by growing more poppies," Ben said. "I'm sure you're now planning to buy even more land to grow more poppies."

Matt and Ralph both nodded.

"Leon Schulz could return to some of his old initiatives and get back to buying more prime land in Florida for the same purpose," Ben said. "There will probably be a big conflict like there was last time."

Another brief silence ensued.

Ralph looked at Ben and nodded slowly. "Your point is well taken."

Matt pursued his lips and glanced at Ralph. "We should do some more weapons training soon. We want to maintain our sharpness on our weapons skills. It looks like the danger could explode for us in the near future."

Ralph gave a thumbs-up.

"We'll continue to help you with that," Ben said.

Chapter 2

In Zurich, Switzerland, the three leaders of the Hidden Empire met to discuss their current status.

"Well, we still have Leon to deal with, but we destroyed the rest of Belzex along with their hi-tech fortress," Zeus said.

Odin shook his head slowly. "The opposite of what we had intended to do."

"It's good to take some time to discuss what happened and try to be a lot smarter as we go forward," Jupiter said. "We don't want to repeat the same mistakes."

Zeus nodded as he glanced at Odin and Jupiter. "Yeah, we need to take sufficient time to discuss things, and we need to remember we had some bad breaks." He paused. "We need to be careful to not change things that don't need changing."

"The death-ray we've been trying to develop still doesn't work," Odin said. "I've been getting regular reports from the team working on it and they think they're making some progress."

"What's the main problem?" Zeus asked.

"It's not precise enough yet," Odin said. "We want to be

able to kill one person in a crowd and not harm the others. That hasn't worked well yet. In our simulations, at times it will kill the target, but sometimes it will also kill some people close to the target."

Zeus and Jupiter each nodded slowly.

"We also have trouble with the death-ray penetrating certain types of material in certain weather conditions." Odin paused. "Bottom line, we're still having trouble perfecting it. If we use it in a real situation, it must be completely reliable."

Zeus shrugged. "So, we made some different plans, and they almost worked to perfection."

Jupiter nodded. "Deciding to use some plastic explosives that imploded worked well. We positioned them in a way where Leon's hi-tech fortress basically fell in on itself. All the people, all the equipment, and all the technology inside the fortress were demolished, and we didn't destroy anything else in the area."

"And our team that planted all the explosives did a great job." Odin gave a thumbs-up. "We used our technology to make them invisible to all the security detection systems and our guys got in, did their job, and got out quickly."

Jupiter grinned. "So, Leon's plan to protect his team and his fortress by being in a heavily populated area didn't work."

There was a brief silence.

"When we executed our new plan, at least we accomplished most of it," Odin said. "We destroyed the hi-tech fortress and everyone within Belzex except for Leon."

Zeus leaned toward Odin and Jupiter. "We need to talk some more about that. How could we not know that Leon wasn't there when we made the decision to destroy everything?"

Odin and Jupiter didn't comment immediately. They seemed to be thinking.

After waiting about five minutes, Zeus leaned forward and regarded both Odin and Jupiter carefully. "Well, how did we screw that up?"

"I've discussed that several times with our technical team that handled that operation," Odin said. "And they keep telling me that Leon was there just before they decided to set off the explosives."

"But he obviously wasn't there," Zeus said.

Both Odin and Jupiter paused for a moment and then shrugged.

"We need to figure that out," Zeus said. "That could be important if we need to do this again."

Odin and Jupiter both nodded.

*****

That night in a private building on Jekyll Island just off the Georgia coastline, Leon Schulz sat at the head of a conference room table.

Ivan Rowan, a Russian billionaire, Chen Wan, a Chinese billionaire, Bolivar Ortiz, a billionaire and a drug cartel leader from Brazil, and Jorge Soto, a billionaire and a drug cartel leader from Mexico sat at the table with Leon and Jake Bolton.

Leon scanned the faces around him. "We've completed a lot of preliminary discussions, is everyone ready to get started on our big project?"

Everyone nodded.

Leon looked at Jake. "Is the Hidden Empire still keeping in contact with you?"

"Yes, and I informed them about this meeting."

Leon narrowed his eyes. "What did they say?"

"They hope you learned a lesson and they'll support your goals as long as you don't interfere with any of their activities or money flow."

"How will I know if I'm interfering with something they're

doing?"

"They told me they'll continue to contact me in the same way as they have been doing, and I'll need to immediately inform you," Jake said.

Leon nodded and showed an expression of satisfaction.

"Like before, they have no problem with your buildup of technology, including all the advanced AI stuff, and they're not uncomfortable with this new group you've put together," Jake said. "They still think the disruptions you'll cause will provide plenty of distractions which will help them maintain the secrecy they need. They're masters of stealth and titans of influence, but they don't want anyone, especially law enforcement, to have the time to study them too closely. A lot of ongoing disruptions involving global events, especially in situations where law enforcement is involved, will benefit them."

Leon leaned toward Jake. "They got uncomfortable before. They might get uncomfortable again as we move forward."

"They emphasized they're okay with you organizing this new group and starting new projects as long as you stop immediately doing something when you're warned about it," Jake said. "They understand everyone in your new group wants to make a lot of money and be highly organized in establishing and running a worldclass global drug trafficking operation." He paused. "They told me that was okay with them."

A brief silence ensued.

"Do we need to decide on a name for this new organization?" Chen asked.

"I thought we could continue to use Belzex," Leon said. "Jake and I talked about it and decided the name could still work."

Chen glanced at Ivan, Bolivar, and Jorge. "I think there might be too much negative baggage that goes with that name."

All three men nodded.

Chen looked back at Leon and Jake.

"Jake and I talked about that," Leon said. "We decided the Hidden Empire is going to know immediately what our new name is and that getting a new name will have no benefits for us in avoiding any conflicts with them."

"What about problems with law enforcement?" "Ivan asked. "They probably have a lot of information about Belzex that could prove useful to them as we go forward."

"Jake and I also discussed that quite a bit," Leon said. "We think they'll discover any new name we develop quickly and transfer any useful information to their new files."

There was a long silence.

Leon could tell all the new partners were deep in thought.

Several of the new partners conferred quietly among themselves.

After about ten minutes, Chen said, "Okay we're all in agreement. Go ahead and use your previous name. At least not changing the name of our organization will make some things easier as we go forward."

Leon and Jake nodded.

The meeting continued for three more hours where goals, strategy, and tactics were discussed.

Chapter 3

Three elderly men sat in a plush lounge area in an elaborate castle located about twenty miles east of Zurich, Switzerland. None of the men owned the castle, a component of the Hidden Empire did.

As members of the Illuminati, a secret society that controlled a lot of world events, this was one of the main places they met when they needed to discuss important things relating to their operation.

None of them knew the real names of the others. Each addressed his cohorts with an assigned codename adapted from the names of the mythical Greek, Roman, and Norse Gods.

The names Zeus, Jupiter, and Odin were reserved for the three most powerful global leaders. The Illuminati organization was super organized and was divided into three major global segments. Each of the global leaders supervised one of the segments.

The oldest man at this meeting, who had the code name of Vulcan, filled his pipe with his favorite almond-blend tobacco. Smoking helped him relax. After lighting his pipe and taking a puff, he looked at the other two men.

Vulcan noted the man who had the code name of Neptune was still in the process of filling his pipe with what Vulcan knew was cherry-blend tobacco. They were both heavy smokers and they each had their favorite tobacco.

Neptune flicked his lighter and positioned the flame. While taking a few puffs to get his pipe going, he glanced toward his team leader, Vulcan.

Apollo, the third man, didn't smoke.

The three men were of different nationalities, but they all were united by a strong passion to support the goals of the Illuminati. Among other things, it was very profitable for them.

Vulcan leaned forward. "There are several things I want to mention. We can decide if any of them need extensive discussion. Zeus contacted me yesterday using the normal method for sending me any messages. He ensured there was no way any could be intercepted."

"Zeus told me the Specter Matrix was in the process of being reworked," Vulcan said. "He wanted to make sure we knew that and reported any irregularities we found that resulted from the activity."

"Were there some problems that needed to be fixed?" Apollo asked.

"Zeus told me it was just some improvements our financial experts recommended," Vulcan said. "Since law enforcement and other entities are getting more sophisticated and using quantum computers, our experts recommended we do this often as a safety valve."

Neptune leaned forward. "I thought it was impossible for anyone to detect the exact path our money takes through the maze we've established using the various global financial institutions?"

"I made that same point, and Zeus told me we're doing it because we don't want to take any chances, no matter how small. He mentioned we use a variety of algorithms at each decision point in the matrix so that we most likely avoid using the exact same path again."

"If it's different each time, why isn't that good enough?" Apollo asked.

"Zeus thinks we need to change the options at each decision point for where the money flows next. That way someone else with a quantum computer won't be able to duplicate the AI decision choices we make," Vulcan said with a slight shrug.

Neptune narrowed his eyes. "That doesn't make any sense to me at all."

"Don't worry about it," Vulcan said. "I'm sure what Zeus is telling our financial experts to do will make the path of any specific money flow we have more undetectable."

Neptune and Apollo both shrugged and didn't comment.

There was a brief silence.

Another reason we're having this meeting is to discuss some information I got last night from Thor," Vulcan said. "He's doing a good job keeping all of us updated on what's going on tactically."

Both Neptune and Apollo leaned slightly forward.

"We don't yet have any new assignments, but Thor told me our bosses are watching things carefully and will notify us if there's any action we need to take," Vulcan said. "We've completed building three new powerplants. The costs are enormous, but we can afford them if the money keeps pouring in from all the different channels we've established." He paused. "That might be one reason Zeus is trying to give our money flow the best protection possible. We need to keep adding new capability to our AI setup, and that is requiring a huge increase in our power usage."

"How many more powerplants are we building?" Apollo asked.

"Four more are under construction, and three additional ones are planned for the near future," Vulcan said. "It looks like we're going to need them."

"What's the schedule for completion of the four powerplants that are being built?" Neptune asked.

"Thor doesn't know yet but he told me he knew the construction process had been expedited so we can provide the necessary power for all our ever-increasing processes as soon as possible," Vulcan said.

"I know we've installed four more quantum computers in our region, and I know that each one requires a large source of energy." Neptune glanced at the two men around him. "We all know we want to maintain our world dominance in AI, and our expenses will continue to be huge to allow us to do that."

"Do either of you know how huge?" Apollo asked.

"Zeus told me last year it was over a trillion dollars a month to properly cover the whole globe," Neptune said. "And we know that's going up fast."

Vulcan nodded. "I'm sure we'll stay informed about the details there. Zeus and Thor will probably both keep us posted on that."

A brief silence ensued.

"While we're together, let's compare notes on a few things. Do either of you know much about what Leon Schulz is doing now?" Neptune asked.

"Only that he's organized a new group of billionaires, and they just had a meeting on Jekyll Island," Vulcan said. "It looks like they're still going to use the same name."

"Belzex Corporation?" Apollo asked.

Vulcan nodded.

"I heard there are six of them counting Jake Bolton," Apollo said.

Vulcan took a long slow puff on his pipe. "That's correct, and I'm sure you noticed they again have a wide geographic scope."

Apollo narrowed his eyes. "And again, they have a lot of talented AI developers. I see them as again becoming a big problem for us."

"We should stay ahead of them in AI with all the new capability we're developing," Vulcan said. "And we already have better surveillance capability than any other organization on the planet."

"At least for now." Apollo leaned toward Vulcan and Neptune. "We must make sure no one catches us."

Vulcan gave a short laugh. "Yep, and I think we're taking all the necessary steps to keep anyone from catching us." He paused. "We've already talked about that. So, I don't think we need any more discussion on that topic."

Apollo and Neptune both nodded.

"I guess our bosses are glad to still have Jake Bolton available to facilitate any necessary communication with Belzex," Neptune said.

Vulcan gave a slight shrug. "Thor didn't comment on that, but I'll bet they are." He paused. "Having a secure method to communicate with anyone who might cause us a problem is always a benefit. We can't afford to make any mistakes. We constantly need to get big things done and we want to remain invisible to the public while we do them."

"Since we need a hugh cash flow from our drug trafficking business, do you think any new approaches Belzex takes for their drug trafficking could create some problems for us?" Apollo asked. "With the rising costs of staying ahead of everyone in AI, we can't afford to have any large channels of funding cut off."

"We have everything organized where none of the drug cartels are causing a problem. In fact, we have an ingenious setup where we get a share of their profits and they don't even know it," Vulcan said. "We won't allow any disruption in our flow of funds and that's where Jake Bolton will come in handy. We know Belzex has set up a lot of shell companies to facilitate their flow of money from their illegal drug business for their regular operations, and for their plans to greatly expand their AI development. The way we have things set up we'll get a part of their profits too. If Belzex causes us a problem in any way, I'm sure our bosses will send Jake Bolton to let them know about it."

Neptune and Apollo both nodded.

“Up to a point, what Belzex will be doing is good for us. Having the world constantly disrupted helps us to maintain our secrecy. Belzex was useful before in that regard and they’ll be useful again.” Vulcan fixed his gaze on both Neptune and Apollo. “I’ll keep you informed. We’ll probably have another meeting soon.”

After a short pause, Apollo asked, “Do we know what Belzex is doing or planning to do with AI?’

“Not yet, but we have a lot of experts looking into it.” Vulcan took a long puff from his pipe as he scanned the faces around him. “Belzex hasn’t completed any new plans yet. Once they do, that will probably be a main subject in our next meeting.”

Chapter 4

In Mexico City, the six members of Belzex gathered in a lounge area of the expansive mansion of Jorge Soto.

Everyone sat in comfortable stuffed chairs in a u-shape formation.

Leon Schulz studied the people in the room for a moment. He and all the other billionaires had large mansions with rooms conducive to conducting their meetings.

In their last meeting on Jekyll Island, they had agreed to continue alternating their meetings between various sites in different countries. They didn't want to be predictable.

Along with Leon, Jake, and Jorge, Ivan Rowan was there from Russia, Chen Wan was there from China, and Bolivar Ortiz was there from Brazil.

Leon looked slowly around the room. "Now let's get down to business. We haven't thoroughly discussed it yet, but I'm sure in the long run we'll all want to gain more global power in addition to a lot more wealth." He paused. "We'll do that by putting a super long-range plan together and by making sure it's foolproof."

Chen narrowed his eyes. "Are we going to be careful to not step on the toes of the Hidden Empire?"

"How do you know about them?" Leon asked.

"China keeps up with everything going on in the world. We have millions of spies in each of the large countries." Chen gave an evil grin.

Leon scanned the faces around him and gave a short explanation about the Hidden Empire for anyone who wasn't aware of them.

"Do you think they'll cause some problems for us?" Boliva asked.

"Not if we're careful," Leon said. "We'll try to make sure we don't clash with anything they're doing."

"If they operate behind the scenes of what's going on around the world, how will we know?" Jorge shrugged and gave Leon a long stare.

"Once Belzex completes its recovery, we'll have a lot of advanced technology ourselves," Leon said. "We'll know a lot about the details of what's going on globally." He paused. "We'll just need to be careful about the decisions we make and be very careful none of our activities are something that would create a new problem for the Hidden Empire."

Everyone nodded slowly and made no comments.

After a short pause, Leon said, "Let's get moving. We have some important matters to address and a lot of coordination to do. We need to prioritize things and get on with our activities."

Chapter 5

Ralph sat at the round conference table in a corner of his office at Tuxtun in Jacksonville along with four other people. Matt, along with Judy Caldwell (Business Analyst), Todd Green (Technology Specialist), and Steve Baxter (Security Chief) attended the meeting Ralph had called.

"I had a long conversation with Ben yesterday and got a lot more information about a lot of issues," Ralph said, as he looked around the table. "I thought it best to discuss some things immediately. With all the potential problems we could have, we must stay well-coordinated on all the important issues as we go forward."

Everyone nodded slowly as they waited for Ralph to continue.

"Ben assured me the Hidden Empire is still out there. Hugo used his advanced technology and planned to destroy them, but they're too elusive, too powerful, and too invisible," Ralph said. "I think it's safe to say they're the ultimate masters of stealth and they're obviously titans of influence. They have the money, the technology, and the skills to influence all the important issues on this planet."

Judy nodded. "After they destroyed Hugo Wagner and avoided suffering any damage from Leon Schulz, I've spent a lot of time trying to find out more about them. I haven't made much progress. They're almost completely invisible."

"But for some reason, the Hidden Empire didn't kill Leon Schulz," Matt said. "He's obviously still out there and we know he's organizing a new group."

"Yeah, that's what Ben spent most of his time discussing with me," Ralph said, "and Ben has a lot more resources than we do to investigate some of the specifics about that. Ben emphasized that Kyle and other FBI resources all work closely with the NSA, Rand, and the CIA. Even with all their technology and skills, they've found very little additional information about the Hidden Empire." Ralph grunted and looked at Judy. "In your investigation, did you find anything useful you can share with us?"

"I found a lot of shell companies that Leon Schulz has a connection to. They all channel money to Belzex, the name his new group of billionaires will continue to use for their activities," Judy said.

"Did you uncover anything new about the money flow of the Hidden Empire?" Matt asked.

"I still don't think they use any shell companies. And I still don't think they need them, but I did find evidence they have some set up and they might be testing them. I found some money trails this time that tangled around in the Channel Islands and the Isle of Man," Judy said. "They can control the activities of all the global monetary organizations to meet their needs, but it looks like they're preparing to use the Channel Islands and the Isle of Man as components in the Specter Matrix if they ever need them."

Matt nodded slowly. "They probably decided they might, at some point, need to include them to make their money flow even harder to detect."

"By continuing to use the full capabilities of the WTO, the

IMF, the World Bank, and the Federal Reserve System, they're able to operate almost invisibly." Judy scanned all the faces around her. "I don't fully understand it yet so I can't explain it, but it looks like they've prepared to be able to merge their transactions into various and assorted transactions from the shell companies."

Todd looked at Judy. "From what you showed me, they have to be using multiple quantum computers to make their Specter Matrix function in the way they want it to."

Judy nodded. "You, Kyle, Dan, and I will continue to work on this with other specialty groups, including one from Rand."

There was a brief silence.

Ralph leaned forward and gave a short laugh. "I now have some understanding about how the Hidden Empire can keep their monetary transactions invisible to the public."

"The International Acceptance Bank of New York that was formed in 1921 no longer exists, but the Hidden Empire uses modern banks to replace what that bank was doing as part of the long-complicated process," Judy said. "It looks like the Hidden Empire has built a financial process that's impossible to for any of us to uncover."

Matt narrowed his eyes. "I'm sure the Hidden Empire has people in high places in all of those financial organizations."

"Remember the World bank is actually a combination of five different banks, and the Hidden Empire probably has their own people in charge of all the major functions," Judy said. "The public doesn't realize how much power they have. It's easy for them to channel money to any individuals or organizations of their choice. As I just mentioned, they don't need to rely on using any shell companies to maintain their secrecy."

Ralph scanned the faces around him and then focused on Judy. "We're all accustomed to dealing with complicated subjects and we can handle a lot of complexity. You've been over some of this before, but would you give us some more details on those five branches of the World Bank?"

"It's a complex system," Judy said. "How much detail do you want?"

Ralph narrowed his eyes. "A lot." He paused. "Like I mentioned, we're all accustomed to dealing with complicated subjects. We need to get sharper on knowing both the big picture along with many of the details for what we're going to be analyzing."

Judy grinned. "Well, I think you're about to get another test of your analytical ability."

Ralph, Matt, Todd, and Steve each gave a thumbs-up.

There was a brief silence.

"Well, let's get to it," Ralph said. He looked at Judy. "Anytime you're ready."

Chapter 6

In Mexico City at Jorge's mansion, Leon looked at the other five men around him and said, "I think the first item of business we should address is that of our expected additional money flow. We're all already rich and powerful, but we all want more, a lot more."

Chen laughed. "Of course. We're all fully aware that's why we've banded together. All of us are rich but we all want to be a lot richer."

"And more powerful too," Ivan added.

Everyone nodded.

Leon narrowed his eyes. "Now we need to make sure we get things setup to support our goals. Once we get everything worked out and in place, we can expect to have more world power, and we can expect regular deposits into our designated accounts from all our shell companies."

"How soon will we start getting those deposits?" Chen asked.

"It should be very soon," Leon said. "And that should continue to be a common occurrence. We should expect a lot of money to flow in constantly from all our sources. I'm thinking each person should average ten million dollars a month."

"What do you base that on?" Bolivar asked. He leaned forward and stared at Leon.

Leon explained.

"It looks like you planned this out well," Jorge said.

Leon nodded. "There's strength in our joining forces and I figured, if we operated effectively, there will be a lot more money available for everyone. It will be a lot more than any one of us could make by himself."

"And we should have a lot more power too," Chen said. He paused. "That is if we handle things thing well in our dealings with the Hidden Empire." He gave Leon an extra-long stare.

"If we play our cards right, we have a lot of potential." Ivan raised his right fist in the air. "I'm anxious to finalize all the details and ensure we can cope well with the Hidden Empire." He scanned the faces around the room. "As we've mentioned several times, we'll need to do that."

Everyone nodded.

"We all want more power and more money." Leon also raised a fist in the air. "I saw the potential benefits in getting us together and here we are. We'll work out the details as we go along. We don't have to plan everything right now." He paused. "We should be meeting once a week to start with and then, once we're fully organized and operating smoothly, we should continue to meet once a month. There will be more issues to constantly resolve."

"Are we going to start diving into some of the details right now?" Ivan asked.

"We'll get started, but remember, there are a lot of them," Leon said. "Everyone will need to exercise patience as we go along." He paused for several moments. "I have the shell companies all set up and functioning, and I have my personal accounts established to receive the money flow."

Leon reached for his laptop computer. "We'll start with everyone else getting your personal accounts set up." He scanned the faces around him and waited for everyone to get their computers ready.

In a few minutes, Leon said, “I’ll give you all the details you need to set everything up.” He leaned forward and started explaining.

Everyone listened intently and asked some questions as they followed the instructions.

Chapter 7

At Tuxtun in Jacksonville, Judy continued with her explanation of the basic structure of the global financial system. She leaned forward and looked at the men around her.

"The five branches of the World Bank are the International Bank for Reconstruction and Development (IBRD), the International Development Association (IDA), the International Finance Corporation (IFC), the Multilateral Investment Guarantee Agency (MIGA), and the International Centre for Settlement of Investment Disputes (ICSID)," Judy said.

"That's a lot to get familiar with," Steve said, "especially for a security guy."

Judy shrugged. "Gaining some familiarity with this subject-matter will help you in some of the investigations you'll probably get involved with."

Some people nodded, but no one commented.

Judy proceeded to give more details about how they functioned.

"That's a lot for me to digest," Steve said. He glanced at Judy. "But you warned us."

Judy smiled.

Ralph looked directly at Steve. "Judy has a good point. It won't hurt you to know a little more about all this. It might occasionally relate directly to our security."

Steve nodded and made no further comments.

"If the Hidden Empire controls all five branches of the World Bank and also controls the Federal Reserve System along with the IMF and the WTO it's easy to see why they don't need any shell companies to facilitate their cashflow," Matt said.

"Yeah, but I think for extra insurance, they've found a way to merge some of their cashflow with that of the shell companies." Judy went on to give a few more details, then she said, "The Hidden Empire's cashflow will be very difficult to ever find any trace of."

Ralph laughed. "That's a no-brainer." He looked directly at Judy. "Keep working on this with Todd and Rand, as well as Kyle and other law enforcement."

Judy gave a thumbs-up.

Matt nodded. "We'll need to know as much as possible about the global money flow if we have any trouble with either Belzex or the Hidden Empire."

"Remember that Belzex does need the shell companies for their cashflow," Judy said. "Todd and I are continuing to investigate that. His knowledge of AI is proving to be a very useful addition to my knowledge of Finance."

Ralph narrowed his eyes and looked at Judy. "Matt and I want to keep up with what you're investigating. Keep us both posted on your progress and let's continue to have frequent group meetings."

Everybody gave a thumbs-up.

Chapter 8

In Geneva, Switzerland, as CEO of a large multinational corporation, Zeus sat in his plush office along with Odin and Jupiter at the round conference table in a corner.

The door was closed, and the room had been swept for bugs, even very sophisticated ones.

"Well, right now everything is going smoothly," Zeus said. "It's good to be able to relax a little."

Odin leaned forward. "I agree, but I don't think our relaxing is going to last long."

Zeus and Jupiter each raised an eyebrow and stared at Odin.

"Belzex is raring their head again. With their new team in place, they're moving forward quickly with fresh global business plans." Odin narrowed his eyes. "If they're successful, I think they'll continue to want more world power. We haven't succeeded in eliminating that particular threat to our continued dominance over world events."

Zeus nodded. "I think Leon Schulz has an uncontrollable urge for power. Remember, he has the same genes as Hugo Wagner did. They're both descendants of Hitler. I don't think he can control his urge any better than Hugo did."

"That's true," Jupiter said, "but it shouldn't cause us a problem we can't handle. We handled it before and we'll handle it again. We'll make sure they know they can get very rich and powerful if they stay in their lane."

Odin tightened his lips for a moment. "Hugo didn't stay in his lane before and we eliminated him. Leon didn't stay in his lane either and we eliminated everyone around him." He paused. "But we've decided to give him another chance."

Zeus narrowed his eyes. "If Leon with his new group starts doing things we don't allow, we'll make sure Jake Bolton delivers a loud and clear message to all of them."

Odin took a deep breath and exhaled sharply. "As they pursue more power, I don't think they'll stay in their lane. I think they're going to create some big problems for us real soon by becoming bolder as they become more powerful."

"Are they doing anything specific yet that you're worried about?" Zeus asked.

"They're making Belzex an even larger global conglomerate than it was before. On the surface they're planning to be a large provider of technology (including AI capability), food, legal drugs, and oil," Odin said. "Even though there are a lot of other conglomerates doing similar things none of them are led by a man with Hitler's genes."

Zeus grunted. "Yeah, we learned that lesson before." He stared at the other two men for an instant. "We shouldn't make the same mistake again."

Jupiter and Odin nodded.

There was a brief silence.

"We'll keep a close eye on their activities," Zeus said. "If they start doing the same things again, we'll pick the optimal time to put them out of business. We'll make sure we do what's best for us."

"I think Belzex will again concentrate on trafficking illegal drugs. There's a lot of money in that. They're going to promote legal drugs like Morphine and Codeine, but they're going to concentrate more on producing Heroin and Opium," Odin said.

"Even though they all have great wealth, I'm sure they're looking to fast-track their increase in revenue which they'll need for their planned AI development, which I'm sure they'll greatly expand."

Jupiter nodded. "Yep, we know firsthand how much money it takes to make AI fully functional for sophisticated global usage. Supplying power where it's needed is unbelievably costly."

There was a brief silence.

"Morphine, Codeine, Heroin, and Opium are all made from poppy straw, aren't they?" Zeus asked.

"Concentrate of poppy straw," Odin said. "It's referred to as CPS and it's every bit legal as long as it's for producing legal drugs. Belzex will use the legal part as cover, but they'll be putting a lot of effort into the illegal part and setting up their global cashflow to ballon into an astronomical amount. That will help them pay for all the things necessary to increase their power and influence around the globe." He went on to explain.

Zeus leaned forward. "We well know how an enormous amount of money can be a big help in gaining an enormous amount of power." He paused for a few seconds. "And, of course, we have inside knowledge of the whole process." He paused again as he scanned the faces around him. "We need to keep a close eye on what they're doing and make sure they don't gain the upper hand on us. That could happen if we aren't very careful."

"If there are any big conflicts with Tuxtun and Belzex competing on buying land, growing poppies, and producing CPS in several different countries, some of their activities could explode into a big problem or us," Jupiter said.

Zeus nodded as he glanced at Jupiter and Odin. "We'll not just keep a close eye on Belzex, we'll keep a close eye on everything going on with Tuxtun and Belzex. We have the technology and the expertise to do that."

Chapter 9

After Judy, Todd, and Steve had left his office when the meeting had ended, Ralph looked across the round conference table at Matt.

"What do you think about the position Tuxtun is in right now?" Ralph asked.

"As we've often done before, we're dealing with a big tangle of potential problems," Matt said. "I think we can handle the situation, but we need to stay ready to cope with any problems that crop up, and it could be a wide variety of them right now."

"Are you concerned about anything in particular?"

Matt hesitated for a few moments and then shrugged. "We have a lot of potential issues. I'm concerned about everything equally at this point. We'll stay on ready and see how things develop."

Ralph nodded slowly.

"How about you? Do you have any specific concerns? Matt asked.

Ralph shrugged. "I'm thinking the same as you. We need to stay prepared to handle a wide variety of things."

Matt leaned slightly forward. "Do you have any guesses as to how soon we'll be contacted by Belzex?"

"Why do you think they'll contact us?"

"According to what Judy mentioned, we're competitors in buying land to grow poppies and all their new people will probably want to know us better so they can size us up."

Ralph nodded slowly and made no comment.

"I think we can expect to be contacted soon," Matt said. "They might even want to buy some of the land we already own. I think we can expect to get a request for some meetings."

Ralph continued to nod slowly. "And I'm certain their production of legal drugs will be a small percentage of their overall business. They'll concentrate on producing Opium and Heroin."

"Yep," Matt said. "We can be certain they'll concentrate on whatever brings them the most money. They're going to need that money to fund their enormous AI activity that Ben mentioned based on the intelligence information he received from the NSA."

Chapter 10

At Jorge's mansion in Mexico City, Leon explained a lot of details and worked with everyone to get all their personal accounts established to receive the huge flow of expected deposits.

When that was finished, Leon said, "Now we need to start buying more land we'll need for growing poppies. We need a much larger supply."

Bolivar leaned forward. "I've already checked into that. I own a lot of land in South America that I bought for growing poppies, and I own some processing plants for producing heroin and opium." He grinned. "They produce morphine and codeine too and the legal products are, of course, all I advertise. I'm in the process of finding out who owns the plots of land around the ones I own."

"Is the land you own premium plots in the best areas?" Leon asked.

Bolivar nodded. "Of course, and that's why I recommend the additional land we buy is in the same areas."

"Is there any more land for sale around your land?" Jorge

asked.

Bolivar shook his head. “That’s a problem and we certainly want to farm the best land.” He paused. “I’ll find out who the owners are, and we’ll contact them.’

“We’ll make them a huge offer they can’t refuse,” Leon said. “It will be a good investment. Our profits will be tremendous."

“What if they do refuse? Bolivar asked.

“Hopefully they won’t,” Leon said. “It will be a great offer. We have a lot of money, and we’ll get it all back over a period of time, plus a huge profit. That’s what great investments are all about.”

“They might be thinking the same as you,” Jorge said. “They might think over a period of time they’ll make a lot more money than the amount you offer them.”

Ivan nodded. “That’s a good point. They might not want to sell for the reason Jorge mentioned.”

“Or for a host of other reasons,” Ivan said.

Chen stared at Leon for a moment. “So, you better be prepared for a rejection.”

Leon paused for a moment. “Yeah, Tuxtun Industries owned the land I wanted to buy before, and Ralph Gibson rejected my offer.”

There was a brief silence.

Leon looked at Bolivar. “Step one is to find out who owns the land around your poppy fields. You go ahead and do that and then we’ll decide our next step.”

Bolivar nodded and stood to leave.

*****

A little more than an hour later, Bolivar returned to the meeting.

“Tuxtun Industries owns all the land adjacent to my land,” Bolivar said. “The CEO of Tuxtun is Ralph Gibsun.”

"I know who he is," Leon said gruffly. "I've had to deal with him before."

There was a brief silence.

"Well, what are you going to do?" Ivan asked.

"Ralph Gibson knows me and my previous meetings with him didn't go well," Leon said. "The rest of you will need to handle the next few steps we need to take to get that land. To be fully successful in our new ventures, we must get it."

No one commented.

Leon looked at Bolivar. "You're the most knowledgeable about the land. You'll need to call Gibson and see if you can schedule a meeting." He glanced at the faces around him. "If we get the meeting set up, we'll decide who should go, and we might need to take a couple of our best assassins.

Everyone remained silent. Some narrowed their eyes as they waited for Leon to say more.

Leon leaned forward and explained.

Chapter 11

In an elaborate castle located about twenty miles east of Zurich, Switzerland, three members of the Hidden Empire met to discuss some new events relating to their operation.

Vulcan, Neptune, and Apollo gathered for their next meeting to discuss the new information.

"I thought we needed to get back together and sort through some of the new things immediately," Vulcan said. "Don't know about you, but I still have an unsettled feeling, and Zeus just made several significant new decisions we need to review."

Neptune and Apollo leaned forward.

"Zeus told me the Specter Matrix is still being reworked," Vulcan said. "And he wants us to continue to report any irregularities we find that we think might result from the activity."

"Are there some new problems that need to be fixed?" Neptune asked.

Vulcan narrowed his eyes. "Zeus told me our financial experts still need to do some more finetuning due to law enforcement and other entities getting a lot more sophisticated with the technology they're using."

"You mentioned before that the financial experts in several law enforcement agencies are using quantum computers, right?" Apollo asked.

Vulcan nodded. "Right, and that's a big reason we have to

make it impossible for anyone to detect the exact path our money takes through the maze we've established using all of the various global financial institutions."

"I guess the law enforcement agencies have more capability now with the quantum computers," Apollo said.

"A lot more capability in a lot of ways." Vulcan paused. "Again, it's critical that none of the payments we make to various important people and organizations around the world will ever be traced."

"Do we know if we're still the ones who have the best technology on the planet?" Apollo asked.

"We'll talk about that in a few minutes, right now I want to remind you and Neptune about one particular capability we have if we need to use it again. " Vulcan said. "Remember, the Russian KGB has access to a lot of nuclear weapons, all types and the Chinese are helping enhance them." Vulcan leaned forward. "Don't forget we have access to all that. We can destroy anyone on the planet in many ways. We can wipe anyone and any organization off the face of the earth if we choose to. We're currently well embedded in every major country around the globe and we intend to stay that way. That's how we'll eventually reach our goal of a one-world government."

Neptune and Apollo both nodded.

"There's no way we'll forget how we disposed of Hugo Wagner," Apollo said.

Neptune continued to nod.

"It's good to make sure our global money flow won't be interfered with." Apollo looked at Vulcan. "But we know some intelligence agencies are continuing to try to analyze it. Do you think that could turn out to be a problem for us?"

"I asked Zeus about that, and he told me he had discussed it with Jupiter and Odin," Vulcan said. "Our technology is more advanced than theirs, however, we're going to watch their activity closely and take any action that's necessary if they try to interfere with something critical to our money flow."

There was a brief silence.

"Zeus also told me our financial group has made some recent enhancements to the procedures and activities of the World Trade Organization over in Geneva and to the Bank for International Settlements over in Basel," Vulcan said. "After some additional close analysis on how our global money flow is working, our group of financial experts decided to make some adjustments to keep the details of our money flow even more hidden."

"Did Zeus give you any specifics?" Apollo asked.

"I asked about that, but he emphasized that he, Jupiter, and Odin never get into any of the minute financial details, it's too complex," Vulcan said. "He did, however, give me some more information." Vulcan paused and took another puff from his pipe. "He mentioned our financial experts were also looking into making some adjustments to the procedures and activities of the Federal Open Market Committee and the Institute of International Finance."

"I'm not familiar with them. What do they do?" Apollo asked.

"They mainly just fill in some gaps in our global money flow," Vulcan said. "I asked that same question myself and I just gave you the answer I received." He smiled and then took a long slow puff on his pipe. "If I receive more information, we'll discuss at least some of it in future meetings if we need to."

A brief silence ensued.

Vulcan leaned forward. "Remember, Hugo had to use a lot of shell companies to keep his money flow well hidden. Our system is a lot better than his ever was, and we don't need the shell companies to help us operate in complete secrecy. Our full system is complex enough by itself."

"I know some financial analysts in the intelligence agencies are still working to fully define the global financial system," Neptune said, and I know the group at Tuxtun is still working with them." He paused. "It looks like they want to fully understand the mechanics of how it all works."

Vulcan leaned forward. "Don't worry. They'll never figure it all out and we'll constantly stay ahead of them. Zeus assured me of that."

"But that analytical activity could turn out to be a problem for us in the long run, don't you think?" Apollo asked.

"If it does, our leaders won't allow it to continue," Vulcan said.

Apollo looked unconvinced.

"Zeus assured me our financial analysts are keeping a close eye on what the combined intelligence agencies and the Tuxtun group are doing but our analysts don't think that group will ever be able to figure it out," Vulcan said.

Apollo nodded slowly.

"It's so complex that even our financial experts have to go back to the original blueprint and then review a lot of the changes we've made over the years to understand things properly," Vulcan said.

Neptune took a puff from his pipe and leaned forward. "It's good we have powerful people in all the right places so we can make any enhancements at any time when we need to do so."

"That's right," Vulcan said. "We're in complete control. We have people in position to make changes when we need them and they're the only people who can make the changes."

"Hugo was able to get some of his people in position to allow him to use parts of our system," Apollo said.

Vulcan took a long puff on his pipe and looked at Apollo. "Yeah but remember he could only use parts of our system. That's why he had to create all the shell companies."

"And also, Hugo's people couldn't make any changes in our system. They were very limited on what they could do." Neptune said. "I think we can relax on that subject and concentrate on other things we need to discuss."

"That's correct," Vulcan said. "Our leaders will keep up with what's going on and I'm sure they'll take any necessary action they need to."

"We still want to constantly make payments to the right influential leaders to ensure important events go the way we need them to, right?" Apollo asked.

Vulcan nodded slowly. "Yes, that's still an essential process, and we need to always have the right amount of money to make our system work properly. Also, Zeus emphasized that our experts use a variety of algorithms at each decision point in the matrix so that we avoid using the exact same path again. We don't want someone else with a quantum computer to be able to duplicate the AI decision choices that were made. He paused. "I'm sure what our financial experts do will make the path of any specific money flow we have more undetectable to other organizations who're using a quantum computer."

There was a brief silence.

Another reason we're having this meeting is to discuss some information I got last night from Thor," Vulcan said. "He's doing a good job keeping all of us updated on what's going on." He leaned slightly forward. "We've completed two more powerplants to provide the necessary energy for all our advanced AI processes. The costs are enormous, but we can afford them if the money keeps pouring in from all the different monetary channels we've established. That might be one reason Zeus is trying to give our money flow the best protection possible. We need to keep adding new capability to our AI setup, and that is requiring a huge increase in our power usage."

"We just finished three new powerplants last year and now we have two more. How many more powerplants are we building?" Apollo asked.

"Probably a lot," Vulcan said. "With the way technology is advancing, it looks like we're going to need them."

"Are these two new ones fully online yet?" Neptune asked.

"Thor didn't know all the details, but he told me he knew the related activity had been expedited so we can provide the necessary power for all our ever-increasing processes as soon as possible," Vulcan said.

"I know we've installed three more quantum computers in my region, and I know that each one requires a large source of energy." Neptune glanced at the two men around him. "We all know we want to maintain our world dominance in AI, and our expenses will continue to be huge to allow us to do that. The Specter Matrix must perform properly to guarantee the necessary money flow, which currently is over a trillion dollars a month to properly cover the whole globe."

Vulcan nodded. "And we all know that's going up."

A brief silence ensued.

"While we're together, let's compare notes on a few things. Let's get back to what Leon Schulz is doing." Neptune scanned the faces around him.

"We know he's organized a new group of billionaires, and they just had a meeting on Jekyll Island," Vulcan said. "And it looks like they're still going to use the same name."

Vulcan took a long slow puff on his pipe. "And they again have a wide geographic scope."

Apollo narrowed his eyes. "Since they have some talented AI developers. I see this new group becoming a big problem for us again. I think we should eliminate all of them before they get fully organized."

"Zeus thinks we'll stay ahead of them in AI with all the new capability we're developing," Vulcan said. "And we have better surveillance capability than any other organization on the planet."

"Maybe for now." Apollo leaned toward Vulcan and Neptune. "But we need to make sure no one catches us."

Vulcan gave a short laugh. "Yep, and I think we're taking all the necessary steps to keep that from happening." He paused. "I don't think we need any more discussion about that particular topic right now."

A brief silence ensued.

"It's good to still have Jake Bolton available to facilitate any necessary communication with Belzex," Neptune said. "Having a secure method to communicate with anyone who might cause us a problem is always a benefit."

"Do you think any new approaches Belzex takes for their drug trafficking could create some problems for us?" Apollo asked.

Vulcan leaned forward. "Remember, we have everything organized at the moment where none of the drug cartels are causing a problem. In fact, we have an ingenious setup where we get a share of their profits, and they don't even know it. The more profit they make the more we'll make."

"We know Belzex has set up a lot of shell companies to facilitate their cashflow to their regular operations," Neptune said, "and they plan to greatly expand their AI development in the near future."

Vulcan nodded. "We'll keep a close eye on them and not let them gain any ground on us as far as technology is concerned. If they make a lot of advancements but always stay a little behind us, that will probably be okay."

Chapter 12

At Tuxtun in Jacksonville, Ralph and Matt sat in the boardroom with Judy, Todd, Steve, Ben, and Kyle.

"After Ben and I had a long discussion last night about everything that's going on, I discussed it with Matt." Ralph slowly scanned the faces around the table and then gestured to Matt.

Matt nodded. "I thought we needed a conversation with all of us present. Considering what's probably coming up, it's good we stay well-coordinated on all the important issues."

Ralph motioned to Ben.

"Bolivar Ortiz is one of the billionaires in Belzex and he owns a large drug trafficking cartel," Ben said. "He has an on-going war with other drug cartels, and he employs a lot of assassins. Matt has already had an encounter with one of them." He paused. "We need to be aware of the possibility of running into more of them real soon."

"We need to know more about him, and I'm in the process of researching that now." Kyle looked at Judy and Todd. "I'll keep in close touch with you as I proceed."

Judy and Todd each gave a thumbs-up.

"Anything you've found out so far that would be good to mention at this time?" Ralph asked.

"Bolivar Ortiz does indeed have a large group of assassins." Kyle glanced at Ralph and then looked at Matt and grinned. "He seems to have a lot of squabbles with rival cartels and that guy you met in the store parking lot was one of his assassins. His name is Pablo Garcia."

"He was definitely the type," Matt said.

Kyle nodded. "I'll continue researching. Pretty soon we'll know all about a lot of the assassins." He paused. "I've also found some other interesting information: Chen Wan owns a large advanced-technology company. In addition to that, Ivan Rowan and Jorge Soto each own a large manufacturing company. That gives them quite a bit of capability."

"It looks like Leon is following the same blueprint as he did last time," Matt said. "Belzex again has great potential for gaining a lot of power all around the globe."

Ben narrowed his eyes. "If Leon uses all his potential, he has a chance to succeed this time. We know he's a descendant of Hitler. It depends on how much power he wants." He focused his gaze on both Ralph and Matt. "Maybe you'll get a reading on that if he contacts you about buying some land."

Ralph and Matt both nodded.

Judy looked at Kyle. "Todd and I will probably be working closely with you for a while."

"For sure," Kyle said. "Things are probably going to get a lot more complicated quickly, and we'll all need to work closely together and apply some expert analysis to always react in the most intelligent way we can."

Chapter 13

In Berlin, Jake Bolton picked up his ringing phone after he had glanced at the caller ID and knew the call was from his boss.

"Have you had anyone paying you a visit from the Hidden Empire?" Leon asked.

"Do you have any reason for asking?"

"Maybe."

There was a brief silence.

"Well don't keep me in suspense," Jake said.

"From what you've told me about your previous experiences, I'm sure we're being watched closely by the Hidden Empire."

"That's likely the case."

"And you've made it clear they have exceptional capability and can eavesdrop on our conversations."

"Yes."

"Well, we've had a lot of conversations about improving the powerbase for Belzex and we have plans for taking action."

Jake remained silent.

"I thought the Hidden Empire might think we're stepping on their toes and want to warn us to back off."

“There have been no visitors yet, but I won’t be surprised now if someone shows up.”

There was a brief silence.

“Well, for now, we’re moving on and we’ll stay on guard,” Leon said. “We want to buy a lot of land for Belzex, and we want it all to be premium land for growing poppies, so I’m going to set up a meeting with Tuxtun as soon as possible.”

“Anything I need to do?”

“Not unless I have trouble setting up a meeting,” Leon replied.

“Keep me posted,” Jake said. He disconnected.

Chapter 14

The next day in Berlin, Leon scanned the faces around him. "I've set up a meeting with Tuxtun, but we have a lot of planning to do before we have our meeting."

"What are you planning to tackle first?" Ivan asked.

Leon hesitated for a moment and then leaned forward. "We have a lot of potential, and I think we need to decide first how much world power we want in the long run. When we get that figured out, we'll go from there."

No one commented Everyone seemed to be waiting for Leon to continue.

Leon regarded the people around him carefully. "We're a team now and whatever we do I'm expecting all of us to share the costs and the benefits equally." He narrowed his eyes. "We'll configure everything so that all the profits flow automatically to Belzex."

No one argued.

Leon looked at Jorge and Ivan. "We'll set some new goals for manufacturing, and we'll plan to create some new products." He looked at Chen. "We'll create a wide range of new AI products for multiple purposes. Some of the AI products will be used in the manufacturing processes. We know the Hidden Empire has very advanced AI products and they're constantly using them. One of our major goals will be to become more capable than they are in that area."

Everyone nodded.

"Do you have some specific things in mind right now?" Bolivar asked.

"As I mentioned, we'll immediately channel our profits from drug trafficking to Belzex automatically." Leon glanced at a ThinkPad he held in his hand. "Any arguments?"

No one commented.

"And, once we have the additional land in full production mode, I expect those profits to be a huge amount," Leon said. "If we meet all the goals we're going to have, we'll need a large new steady flow of money." He paused and studied his ThinkPad for a few moments. "I recommend we develop the necessary technology to transport our illegal drug products to the various markets quicker and easier. We need new AI products to support better intelligence and better manufacturing processes."

Again, no one commented. Everyone was obviously waiting for Leon to continue.

"We all know we're going to need a lot of new energy to fuel our huge increase in AI usage," Leon said. "We need to complete some new powerplants as soon as possible."

Ivan pressed his lips together and gave a slow nod. "You've obviously done a lot of thinking about this, tell us more details."

Leon spent the next hour explaining his plan.

Chen laughed. "If we can do that it will be quite amazing. We'll definitely be a threat to the Hidden Empire, and they'll probably try to stop us."

"I think we can gain the upper hand in our dealings with the Hidden Empire." Leon continued to explain some more things he had in mind.

"If we can do all that, we'll be the most powerful people on earth," Jorge said.

"Now you're getting the point." Leon gave an evil grin. "That, of course, is our ultimate goal."

Ivan leaned forward. "We can't be sure how powerful the Hidden Empire is right now." He paused. "They were damned powerful before and they're probably constantly improving".

"I've done a lot of thinking about that." Leon narrowed his eyes. "We have the ability to set up a better operation than Hugo did, and a better operation than I had before. Jake and I have a lot more experience with the Hidden Empire , and we'll be smarter than we were before."

No one commented.

Leon proceeded to give a few more details.

"So, you're thinking we're going to try to conquer the world sooner rather than later?" Ivan asked.

"We don't want to wait too long," Leon said, "but, we don't have to make that decision now. We'll have a lot of options as things develop. We'll be ready to take advantage of any opportunities that can help us. We'll steadily develop an enormous amount of power so that we can do that."

Everyone nodded.

"We'll be very careful as we go forward and we'll use a lot of stealth," Leon said. "We don't want what we're doing to be obvious to anyone."

"I'm sure the Hidden Empire will know what we're doing, and we'll probably get a warning like Hugo did and like you've gotten before," Chen said. "I think it's very likely they'll continue to use Jake as the messenger."

Leon nodded. "That likely will be the method they'll use."

Ivan leaned forward and stared at Leon. "I hope you don't ignore the warnings like Hugo did."

"Hugo got a lot of warnings, and he ignored all of them," Leon said. "I think we'll have a little slack, so we won't stop what we're doing the first time we get a warning." He paused. "We'll stay very alert and make sure we don't slow down our momentum. We'll continue to concentrate on making fast progress to gain the upper hand."

Chen shook his head slowly. “We’re going to be treading on dangerous ground.” He looked directly at Leon. “We’ll need to be extremely careful.”

“Like I mentioned, we’ll be careful, for sure,” Leon said. He paused for several moments. “You can be a great help here by making big improvements in our AI capability so we can monitor more world activity and maybe even gain some new useful knowledge about the Hidden Empire.” He narrowed his eyes and stared at Chen.

After a short pause, Chen nodded.

Everyone raised a fist into the air.

Leon looked directly at Ivan. “Do you still have any close friends in the KGB?”

Ivan nodded. “I still have a few.”

“This might be a good time to check with them to see if any alerts are being put out by the Hidden Empire,” Leon said.

Ivan gave a thumbs-up.

Chapter 15

In a private conference room at the headquarters of a large international conglomerate in New York City, Zeus, Odin, and Jupiter began their next meeting.

Zeus leaned forward. "We have a lot of people watching Belzex carefully. It's obvious Belzex wants more power. They're making big plans. If they give us any real trouble, we have a lot of options. The Russian KGB has access to many weapons, all types, and therefore we have access to them too. We'll be ready to take any action we need to take."

"We can easily destroy Belzex if they prove to be a big problem for us," Odin said. "It's good Jake Bolton is still around and working for Leon Schulz. One of our messengers will contact him when necessary."

"Is there anything going on right now that has you concerned?" Jupiter asked.

Zeus nodded. "Quite a bit." He went on to explain.

"There's no doubt that Leon has some things in mind that are much bigger than just making a lot more money," Odin said as he looked directly at Zeus. "Do you think Leon is trying to duplicate the AI capability that we have?"

"I think he's not just trying to duplicate, but he's trying to have more capability than we have," Zeus said.

"If he does somehow get more capability than we have, will that be a problem to us if he doesn't plan to conquer the world?" Jupiter asked.

"Not necessarily. We like all the disruption he'll cause," Vulcan said. "A lot of disruption in the world makes it easier for us to keep all of our activities hidden." He paused. "However, if he decides to conquer the world, we'll be in his way, and he'll try to eliminate us."

There was a long silence.

"How will we know if he decides to conquer the world?" Odin asked.

"We'll need to constantly monitor all Belzex's plans. We can do it. Over the years, we've amassed a huge amount of capability," Zeus said. "Each of us has a powerful regional technology center, and each technology center focuses on different priorities. We have a lot of planned overlap for backup purposes, but there are also some specialties each site has."

Jupiter and Odin both nodded.

Zeus grunted. "We must be powerful and we must use that power when other people show up who also have a tremendous amount of power."

"It's understandable why Hugo felt he was the most powerful entity on the planet," Odin said. "Maybe Leon has learned his lesson and will stay in his lane this time."

Zeus nodded. "We'll stay on alert and make sure Belzex is eliminated if they get out of their lane." He paused. "We'll know as soon as it looks like they've become a problem. I'm totally confident we'll be able to keep up with all the plans Belzex is making. We're constantly making improvements to our AI. We have more capability right now than we did when we completely wiped Hugo's operation off the map and eliminated Leon's previous hi-tech capability."

Odin narrowed his eyes. "I'm still very concerned about what Belzex is doing now. Leon has resurrected them from the ashes and reorganized the whole group." He paused. "And Belzex is again trying to set up a joint financial office in the United States. This time it's in Jacksonville, Florida."

Zeus narrowed his eyes. "We need to stay on top of that activity."

"Our advanced technology teams are constantly making improvements to our overall capability, and we'll ensue nothing catches us by surprise," Odin said.

Zeus and Jupiter gave a thumbs-up.

There was a brief silence.

Zeus leaned forward. "The state of AI capability is a moving target, and we need to stay ahead of the crowd."

Odin nodded. "All the American intelligence agencies are improving their technology, especially the technology they use for surveillance."

"We need to do more to keep phone calls, emails, texts, and other similar forms of communication from being intercepted," Jupiter said.

Zeus and Odin nodded.

"We'll continue to make good use of the Specter Matrix to hide all our funding activity." Zeus made a fist with his right hand and raised it high into the air. "We'll make sure we keep the upper hand on all the aspects of advanced technology as we go forward."

Jupiter and Odin each raised his fist in the same manner as Zeus.

Chapter 16

The next morning at Tuxtun in Jacksonville, Matt and Ralph sat in the boardroom with Judy, Todd, Steve and three other people.

Justin Mason, the Jacksonville Police Chief was there. Ben Fulton and Kyle Rankin from the FBI rounded out the group.

"I've been informed that elements of the World Bank, the International Monetary Fund (IMF), and the World Trade Organization (WTO) have again established some type of joint office in the Jacksonville area" Ben scanned the faces around the table. "As usual, there's no direct information about who funds this office, but our financial analysts in the FBI tell me none of the money comes directly from any part of the World Bank, the IMF, or the WTO."

"How sure are they about that?" Ralph asked.

"They can't be absolutely sure, but they're also getting an opinion from the CIA." Ben gave a slight shrug. "The CIA has used all their normal methos for trying to check the funds, and nothing has appeared that has any connection to the World Bank, the IMF, or the WTO."

"I'm guessing it's Belzex. I think they keep trying to operate a joint financial office in the United States because the

Federal Reserve is here, and we know they're a big part of what's going on with financial operations throughout the world," Matt said.

Kyle nodded. "It's a big benefit to be able to have face-to-face meetings on short notice."

"Is Tuxtun still expected to play a big part in investigating and confronting all this?" Judy asked. She looked directly at Ben.

Ben grunted. "Yeah, you've been a big help in the past and we hope you'll continue to be. Most of our top officials who're involved in any aspect of this overall situation want to keep you involved."

"There's probably a lot more involved here than we know about," Kyle said. "The Hidden Empire might be planning to support Belzex in some of their activities, and we know you're going to be dealing with Belzex soon on the issues involved with the additional land you bought for growing the poppy crops for expanding your legal drug business."

"This whole thing is turning out to be a big, twisted puzzle right at the beginning." Ralph leaned forward. "We'll get it resolved like we've done before. However, as soon as possible, I'd like to get back to just running a business like we used to do."

Ben shrugged. "I think those days are gone." His gaze took in both Ralph and Matt. "A lot of global issues are popping up that affect all of us, especially Tuxtun. We need your help, and my guess is that you're going to need our help too as you go forward in dealing with Belzex." He paused. "I think you'll need our help one way or the other."

"What do you mean by that?" Ralph asked.

"If you don't sell them the land they want, they won't like it and they'll turn up the pressure on you," Ben said. "If you do sell them some land, they'll want more and will continue to pressure you in various ways. I'm sure they won't hesitate to kill you if they feel they need to do that to help them meet their goals."

Matt looked at Ben. "Are you still going to help us refine our skills to defend ourselves? It's looking more and more like we need to continue doing that."

"I'm ready to make reservations for you at our new weapons training facility as soon as you have a half day to spare," Ben said.

"Finding a half day to spare is going to be a little difficult right now," Ralph said. "Running our business is requiring more time due to all the new things we're doing." He paused. "However, we want to stay alive, so we'll find the time."

"Steve, Judy, and Todd should get some training too," Ben said. "They'll also have more threats to contend with." He looked at Ralph and Matt. "Their training won't be nearly as rigorous as yours, but it will be enough to help them protect themselves better than they can right now."

Everyone gave a thumbs-up.

Chapter 17

A few days later in Berlin, Leon Schulz gathered with the usual Belzex group in a spacious lounge in his elaborate home.

"Well, we've made some real progress in multiple areas in the last couple of days," Leon said. "We now have a meeting scheduled with Ralph Gibson and Matt Baker at Tuxtun next week. It was quite an ordeal to get them to agree to a meeting after the problems with the last one." He paused. "It took a lot of effort to convince them that I hadn't planned to knock them off during the last meeting we had scheduled to discuss buying the land. I even had to agree to a weapons check before entering their headquarters building."

Everyone nodded and made no comments. They were obviously waiting for Leon to continue.

"Now let's get down to business." Leon scanned the faces around him. "We know Tuxtun now owns over four thousand acres of land in Columbia." He looked directly at Bolivar. "Do you have any opinion about how much of that we need to buy?"

"We should try to get all of it within the next year. Six hundred acres would be enough to start with. That alone would be enough to expand our production of heroin and opium sufficiently for the next several years."

""Yeah, six hundred acres should be a good amount to start with. No need to shock them with too big of a number," Ivan said. "We can always offer to buy more when we think the time is right."

No one argued.

A brief silence ensued.

Leon turned toward Chen. "You have a large advanced-technology company that produces a lot of AI products. We need to design and produce some super advanced AI products to help Belzex move forward toward our goals faster. Can you help us do that do that?"

"Do you have some specific products you're thinking about?" Chen asked.

"Belzex needs some new products that will help us be more dominant in the world business arena," Leon said. "I not only want us to be a more powerful force in the world, but I also want us to become an even more powerful force in the business arena."

"Is conquering the world still your primary focus?" Ivan asked.

"In the long run," Leon said. "Remember going too fast didn't work out well for Hugo Wagner, and it didn't work out too well for me last time either."

Chen leaned toward Leon. "Exactly how fast do you want to go?"

Leon shrugged. "We certainly want to wield a lot of power soon all around the world by utilizing all our capabilities. I want to design and develop another super-powerful control center real soon, and I want it to be even better than any we've had previously."

Ivan narrowed his eyes. "Like you mentioned, it didn't work out well for Hugo or you before."

"He didn't handle everything in the proper manner," Leon said. "And, last time, I didn't either. We've all been in too big of a rush."

No one commented.

"We know the Hidden Empire is keeping a close eye on us, and we'll keep that in mind as we move forward in creating a more dominant position in the world for us," Leon said. "We'll be a lot smarter than Hugo was, and I'll be a lot smarted than I was last time."

Everyone nodded slowly and again no one commented.

Leon gestured toward Ivan and Jorge. "Each of you owns a large global manufacturing company. Can you make sure any super AI products Chen produces are manufactured quickly as we need them?"

Ivan and Jorge both nodded.

Leon looked back at Chen. "We'll all work together to produce the outstanding results we need."

Everyone gave a thumbs-up, Bolivar and Jake included.

There was a brief silence.

"I think you've come up with a great initiative, but I don't think the Hidden Empire will be comfortable with what we're doing," Chen said.

Leon grunted. "I think you're right and I'm guessing Jake will be contacted very soon to deliver an urgent message to us."

Jake nodded and made no comment.

Leon jerked a thumb toward Jake. "He has given me all the details about the times he was contacted when he worked for Hugo, and I know the details since he's been working for me." Leon gave a short laugh. "Anyway, we'll decide on an appropriate message for Jake to give them that will make us seem less threating. We'll emphasize we just want to be dominant in global business."

"At least that part will be true," Ivan said.

Jorge, Bolivar, Jake, and Chen all nodded.

"Let's work out some more details on how we can work together to produce everything we need to go forward," Leon said. He waved everyone over to a large table.

Chapter 18

Late the next day, four elderly men sat in cushioned chairs inside an ancient but elaborately furnished mansion in Rome, Italy.

Vulcan and Neptune each filled his pipe with tobacco.

Zeus knew their habits and he knew smoking helped them relax. He didn't smoke and he knew Apollo didn't smoke but he patiently waited for Vulcan and Neptune to get ready to give their full attention to what he was going to tell them. He needed to make sure they were fully informed about some critical new items.

Vulcan lit his pipe and took a few puffs.

Neptune was a little slower. He flicked his lighter and positioned the flame. After taking a few puffs to get his pipe going, he glanced toward Zeus.

Zeus scanned the faces of the three men around him. "We're masters of stealth and titans of influence, and we want to always keep it that way." He paused. "We need to increase our level of global money flow, and we'll, of course, use the Specter Matrix to do that." He paused. "Some issues have emerged that we need to discuss immediately." He studied the faces around him for a moment. "That's why I called this meeting." He paused again and narrowed his eyes. "I think it's smart to continue to meet here for our regional discussions. Jupiter and Odin are doing the same thing in their regions. We've always been well organized and that has preserved our immense power from the beginning of our existence."

The other three men nodded slowly.

Zeus leaned forward. "We're keeping a close eye on Belzex. They've just finished some meetings in Berlin. Most of their meetings have been face-to-face and in different places around the globe so far."

"That's smart," Apollo said. "That's a lot like we do, and it works well for maintaining secrecy." He glanced at Vulcan and Neptune.

Vulcan and Neptune nodded and then each took a big puff on his pipe and exhaled slowly. They obviously were waiting for Zeus to continue.

Zeus looked at Vulcan. "You're the leader of your group. What do you think?"

"What you mentioned is very interesting. Have you been able to listen to any of Belzex's conversations in any of their meetings?" Vulcan asked.

"We have a lot of our experts working on it," Zeus said. "They've just developed some excellent technology to protect their privacy. It's surprisingly advanced. We're working to break through the technological barriers they've installed. We don't think it will take us much longer."

Neptune rubbed his chin and stared at Zeus. "I thought our technology was superior to any in the world."

"It is," Zeus said. "I think it will be just a matter of hours before we'll be able to resume our normal surveillance. As I mentioned, the technology they're using is surprisingly advanced. We haven't yet had time to finetune our technology to zero in on any of their ultra protected locations to pick up conversations inside a specially inside a specially constructed building."

Vulcan leaned toward Zeus. "And you think it will just take our experts just a few more hours to get this worked out?"

"That's what our experts have told us, and they've always been very reliable," Zeus said. "Even though the new phones Belzex has developed are extremely sophisticated with amazing, advanced technology, our experts assure me we'll be able to intercept their phone conversations again within a few more hours, even if they're inside their specially constructed buildings and rooms."

Everyone nodded as they focused on Zeus.

"Trust me," Zeus said. He rubbed his chin slowly for a few seconds and then studied the three people sitting around him. "I've been in our organization for a lot longer than any of you and I thought it would be good to take a little time and make sure you know some important things about our history," He paused. "We're all aware that, even though we call ourselves the illuminati, a lot of other people refer to us as the Hidden Empire."

Vulcan, Neptune, and Apollo all nodded and waited for Zeus to continue.

"You all know we've existed for a long time, and I want to be sure you know some important facts," Zeus said. "Jupiter, Odin, and I have discussed this before, and we all think it's a good idea to take the time at some point to make sure everyone in our organization is aware of a major aspect of our history." He leaned forward. "I think this is a good time to do exactly that."

Everyone waited for Zeus to continue.

"All of you already know many things about what I'm going to tell you, but I want to ensure you're keenly aware of some events that occurred at a certain time in our history. It's important for everyone in our organization to know most of the things about our past as we go forward." Zeus gave a slight grin. "Odin, Jupiter and I agreed that we need to continue to take the time to make sure all our members get the necessary history lessons, and I especially want members in my region of responsibility to know more relevant facts that could prove to be very valuable at certain times."

Vulcan, Neptune, and Apollo nodded.

"Even though we've talked about some of this before, it will take a long time to make sure everyone has all the relevant facts they might need at given points in time," Zeus said. "So, Jupiter, Odin, and I decided to continue this process by making it a part of each of our future meetings."

The other three men continued to nod.

Zeus leaned forward. "I've mentioned this previously, but it bears repeating, the Knights Templar organization contained the most famous heroes of the Crusades. Years later when the Pope abolished them, they joined the Free Masons. As things evolved the inner circle of the new combined group formed our organization and we became known as the Illuminati. Another name for the Illuminati is The Enlightened Ones." He paused. "And eventually, we still want a one-world government and a religion that controls the population." He scanned the faces around him. "That's the ultimate goal, but we're not rushing toward that. We'll eventually abolish private property, but we're moving forward very slowly on purpose. And remember, we have some highly prestigious global groups controlling the pace on that."

Vulcan, Neptune, and Apollo each gave a thumbs-up.

"We must keep large amounts of money flowing in the Specter Matrix to properly fund all the individuals and groups who have a lot of influence over important things we need to get accomplished," Zeus said.

"You've mentioned several groups in the past that are active in forming some aspects of our long-range plans and goals," Vulcan said. He went on to mention some of the names.

Zeus nodded. "And remember, Odin, Jupiter, and I are active members of all these groups."

"I remember some of these groups have meetings around the world that are closed to the public. And at least one of these groups has an annual, unofficial, invitation-only conference of over 100 people from North America and Western Europe," Neptune said. "About a third of them are from government and the rest are from finance, industry, and education."

"That's correct." Zeus paused. "And remember, we influence these groups to come up with action plans. That's one way we control the pace of the changes we make."

There was a brief silence.

"Are you willing to comment on the current action plans?" Apollo asked.

"The details sometime get very complicated," Zeus said. "Think about the tasks you've been given over the last few years. Those are good examples of various aspects of our action plans."

"And we know you must modify your plans at times," Vulcan took a quick puff on his pipe, "like when you destroyed Hugo Wagner and his operation."

Zeus nodded. "Taking the necessary actions for the protection of our organization are made easier because we control the Russian KGB. When we need them they provide hit squads and heavy weapons, like the missiles."

Vulcan, Neptune, and Apollo all nodded slowly.

"We need to be very careful to make sure nothing disrupts our ability to maintain our flow of money which we must have to continue to operate. We must constantly pay all the right people off. And we must pay a lot of them big bucks to steer some important world events in the direction we need them to go," Zeus said.

"So, even doing a lot of drastic things is necessary when we must maintain the money-flow we need to continue to exist, right?" Neptune asked.

"Right," Zeus said. "Over the years we've had to do a lot of drastic things, and we'll need to continue to be in position to do so."

"Do you think we'll need to eliminate the current version of Belzex?" Neptune asked.

"Maybe," Zeus said. "And we'll continue to maintain close tabs on them." He paused. "We must make sure we can influence powerful leaders who are making the important decisions in politics, technology, business, and banking. We must maintain a power structure that can ultimately create a new world-order."

A brief silence ensued.

"Technology is becoming more and more important in all of our activities," Zeus said. "If we can't ensure the Specter Matrix functions properly in delivering the right amount of money to the right accounts without the digital path being traced, then we have big problems. We need to always be in position to take drastic actions when needed."

Apollo narrowed his eyes. "Thanks for reminding us of some of the reasons drastic actions need be taken in order to maintain our power."

"The Hidden Empire must always be able to control important events in the world. Jupiter, Odin, and I will keep our members fully informed and fully aware of the reason drastic measures need to be taken at times," Zeus said.

Vulcan, Neptune, and Apollo all gave a thumbs-up.

Chapter 19

Ben, Matt, Judy, Todd, and Steve joined Ralph at the round conference table in his office.

"We're dealing with a lot of odd things right now," Ben said. "I thought it would be beneficial to huddle and compare notes so we can remain as coordinated as possible as we collectively pursue a wide range of subject matter."

Ralph nodded and looked at Judy. "You're our financial expert, and you've been taking a close look at the financial picture at Belzex. I think that would be a good place to start this meeting."

"I'm a firm believer in following the money," Judy said as she scanned the faces around her. "It looks like Belzex is funneling an even higher percentage of their money through shell corporations now. Multiple new accounts have been established in Bermuda, Switzerland, the British Virgin Islands, the Canary Islands, and Singapore."

Ben laughed. "That's quite a big step. They must be planning on doing a lot of business." He looked at Ralph. "Have they contacted you Yet?"

"We have a meeting scheduled in a few more days, here," Ralph said. "They had to agree to a weapons check before they entered the building."

"What about all of you?" Ben asked. "Are you going to be carrying weapons?"

"Of course," Ralph said. "We don't want to take any chances, and we'll be prepared for anything."

There was a brief silence.

Ben tightened his lips, nodded slowly, and looked at Judy. "During all of your research did you find out any more about the Hidden Empire?"

"Not really and I probably won't. They're very well secluded and they don't need shell companies," Judy said. "During my latest research, I got no hint of there being a Specter Matrix. Todd and I will continue our research on them, however; the process is complex. We know they want to keep every aspect of their funding well hidden, and apparently they're doing that."

"I'm glad the NSA has quantum computers and can help us out. They have some new information about the Specter Matix which I'll share with you in a few minutes." Ben leaned forward. "We already know the Hidden Empire is composed of a very secretive global group of power-elites who are directly involved in the money flow, and they essentially control the major economic activities in the world." He glanced at Ralph and Matt. "We might be having to deal with them too before we finish investigating Belzex."

Ralph and Matt nodded and didn't comment.

No one else commented either.

Ben kept his gaze focused on Ralph and Matt. "From what

you mentioned earlier, obviously you're prepared to handle any violence from Belzex if it erupts at the meeting you're having about buying the land."

The Tuxtun group all nodded.

"We won't take anything for granted," Ralph said. "We'll be prepared for anything."

Ben nodded slowly and then paused for a moment. "That joint financial office that's being established here in the Jacksonville area involving The World Bank, the IMF, and the WTO." He scratched his head and leaned forward. "We've joined with the CIA and have jointly put some of our best financial analysts on the investigation. Everyone on the team agrees that none of the money for establishing it comes directly from any part of the World Bank, the IMF, or the WTO. However, they discovered that Belzex is involved again."

Judy nodded. "That was mine and Todd's guess."

"Yeah," Ben said. "We have a complicated financial situation involving that office, and we need to solve some more of the puzzle as soon as we can."

"Is Kyle going to continue to work with Todd and Judy to investigate it?" Matt asked.

"Of course," Ben said. "Quantum computing along with some complex aspects of the AI issue have emerged as a big part of the puzzle surrounding it. We have a lot of technical issues to resolve. We'll all need to stay closely coordinated as we go forward."

There was a brief silence.

"Like we mentioned: since Leon Schulz and Hugo Wagner each set up a similar one before, it's probably for similar reasons," Judy said. "Kyle, Todd, and I uncovered some interesting details that need to be further analyzed. We'll continue to work hard to find all the answers and we'll keep everyone posted."

Ben looked at Todd and Judy. "I recommend the two of you have daily meetings with the joint CIA and FBI team that's investigating this as you go forward. I think it will be better for all of you to constantly compare notes."

Todd and Judy both nodded.

"We've known for a long while that various aspects of all the global monetary organizations are used by some powerful groups to secretly funnel money to activities and organizations they want to support," Ben said. "Having an office in the United States with specialists from each of the global organizations makes things easier to do that in a lot of ways."

"It's probably a given that one of the powerful groups you alluded to is the Hidden Empire, right?" Judy asked.

Ben narrowed his eyes. "Yeah and we know Belzex is involved in some way and maybe even working with the Hidden Empire. This is something we all need to gain more insight into."

Judy glanced at Ralph.

Ralph gave a thumbs-up. "I fully support Judy spending a lot more of her time on this."

Matt looked at Judy. "You mentioned you and Todd had found some interesting details you need to investigate further. Care to mention any of them?"

After giving a slight shrug, Judy said, "Sure as long as you remember we haven't done much investigation yet and our knowledge will be very limited."

Matt gestured to Judy. "Go ahead."

"We need to remember that our Federal Reserve is also part of the overall global money flow and the people who established the office here probably intend to have a lot of face-to-face meetings to avoid their phone calls being intercepted by the National Security Agency," Judy said. "Like they probably decided the times before, having an office over here in the United States makes things a lot easier, especially in dealing with the Federal Reserve."

Matt nodded. "I'm sure that's the case. We know NSA has improved their technology and are continuing to do so. Phone calls can now be intercepted no matter how good their encryption is."

"If any aspects of this new office prove to be directly connected to the Hidden Empire this could end up getting very complicated for us," Ralph said. "We're pretty sure Belzex is involved."

Ben grunted. "Yeah, we know Leon Schulz is also a descendant of Hitler, so he probably has a lot of urges that are hard to resist. He might even be working with the Hidden Empire this time." He paused. "We need to keep this in mind as we go forward. This could get very complex, but I think we can handle it."

Everyone nodded.

"I think we can be certain Belzex will soon have advanced AI if they don't already," Judy said. "We know they were pretty advanced before they were broken up."

Ralph grunted. "Now they're back and reorganized. It's logical to think Leon Schulz still has ambitions of being dominant on the world stage.

"We know they have enough money to afford a super technological setup with the best artificial intelligence, spy satellites, and such," Ben said. "This situation will probably continue to get more complex and more difficult to untangle as we go forward. I think time is of the essence here."

Todd narrowed his eyes. "We know our Federal Reserve is probably involved in this. The owners of the Federal Reserve have always been masters of disguise as it applies to what they're really doing. They have a long history of covering things up."

"We already know a lot about what you're probably going to say, so don't bother giving us another history lesson," Ralph said.

Todd shrugged as he scanned the faces around the table. "Just wanted to keep everyone reminded about the depth of what we're dealing with."

"With the exception of the Hidden Empire, I think we're all well aware of who we're dealing with," Ralph said. " And we're beginning to get a pretty good perspective on them." He looked directly at Todd.

Todd didn't comment as he gave another shrug.

"Many people connected directly to the Federal Reserve have been ruthless about things over the years and they're probably still that way," Ben said. "We know many descendants of the founders are still in place."

"And don't forget we'll likely be dealing with the Hidden Empire on many aspects of this as we go forward, no matter what subject matter we choose to concentrate on," Matt said. He glanced at Todd and Judy. "We need to improve our intelligence gathering related to about everything we have any contact with."

"The Hidden Empire continues to add a whole new dimension to the situation," Ben said. "We must learn more about them and get a lot of things figured out real soon and then take the necessary actions we need to take, within reason, of course."

Everyone gave a thumbs-up.

"And speaking of that, I think we have some interesting information to think about right now." Judy leaned forward. "Two specific banks, that do what the International Acceptance Bank of New York used to do, had a record number of transactions last week. Something big is going on and we need to figure out what it is before it's too late."

"Why do you think that?" Matt asked.

Judy looked at Matt and kept her gaze locked on him. "You know I'm a big proponent of following the money, and that's exactly what I've been doing." She paused. "Using an acceptance bank to deliver money to the right people creates another layer of secrecy. My guess is that a lot of money has just been delivered and the people who are sending it want the transactions to be as inconspicuous as possible."

"You've told us that before, but what goes on there is still a bit vague to me," Ben said. "Will you explain acceptance banking again for us?" He narrowed his eyes and leaned toward Judy.

Judy gave a slight nod and paused for a moment. Everyone was staring at her.

"I've mentioned before that the way it works in many cases is complicated and difficult to explain in a simple fashion, but I'll give it a try again." Judy pressed her lips together, took a deep breath, and exhaled slowly. "Acceptance banking is basically a process of guarantying repayments of loans." She scanned the faces around the table. "After studying it closely for a while longer, I'm more convinced than ever that it's another good way of delivering money to people who you want to remain in the background as much as possible."

There was a brief silence.

"In my opinion, that's not very clear," Ben said. He shrugged. "Sorry, but I need a few more details about how it works, and why it's a good method to use."

Judy gave a slow nod. "I mentioned it could be a complicated process, and I tried to summarize." She hesitated for a few seconds. "Maybe a little more background will help make some things a little clearer."

Everyone kept their gaze on Judy and leaned forward.

"Our Federal Reserve System was created in 1913," Judy said. After a few years of creating money out of nothing, the founders of the Federal Reserve System decided they needed this bank. One of the founders took the initiative and created the International Acceptance Bank of New York in 1921."

Ben nodded. "I remember you saying something about that about a year ago, but I've never quite understood the relevance of the various aspects of the overall process."

Judy paused for several moments and studied Ben. She leaned forward. "Okay, I get your drift."

Ben smiled. "I just need to understand a little more about how the banks you referenced are accomplishing their objective of getting the money to people in a way that allows them to not become visible in the process."

Judy nodded. "I'll get to that, and remember, as more time passed, the powers behind all the global money flow found out they also needed the World Bank, the IMF, and the WTO. These institutions were added to the global monetary system much later." She shrugged. "But let's not get off the main subject. Let's get back to thinking about why the two banks I referenced just had a huge increase in transactions." She spent about thirty minutes explaining more of the details.

A brief silence ensued.

"I still don't understand the connection between the increase in transactions and what we're concerned about regarding Belzex," Steve said, "and maybe regarding the Hidden Empire too."

"Let's just say this additional process in the global money flow further legitimizes things and helps move large amounts of money to the intended destinations, organizations and people." Judy gestured toward Todd. "He and I both think we need to be concerned. A lot of people in the shadows might have just been paid off and many of those people could be assassins."

Todd nodded. "Obviously the people making the payments don't want the money trail to be discovered, but Judy and I keep a sharp lookout for any big changes in any element of the global money flow."

Ralph scanned the faces around the table. "I'm glad we're having this discussion. I'm beginning to more clearly understand the issues here."

Ben nodded. "We'll benefit from this knowledge in the future as we move forward with all our global business issues, including the financial part. I think it will be good to go a little deeper on this general subject before we stop."

"I'd like more details as to how the Hidden Empire is funded." Ralph gestured toward Judy and Todd. "Go ahead and give us some more information."

"I think any additional details I give you right now will make things even more complicated," Judy said. "You need a chance to digest what we've just been talking about."

Ralph shrugged. "Go ahead and give it a try. We can all handle it."

"Okay," Judy said. "You're in charge." She smiled as she extended her arm on each side with the palm up.

Judy hesitated for a moment and then continued.

"Todd and I have uncovered many reasons why so many

worldwide payments remain hidden," Judy said. "Earlier, the more Todd and I dug into this the more entangled everything got. We've worked hard to untangle things."

Todd nodded. "We got some more training recently by trying to trace the money Leon Schulz was paying various people and organizations to help further his objectives, but it was nothing like what we've run into by trying to gain some insight into how the Hidden Empire's money flow works."

"Just summarize it for us," Ralph said. "We don't need all the details."

Todd gestured to Judy.

"We know the Hidden Empire has a system of paying organizations and people by using what they call their Specter Matrix," Judy said. "We also know the Hidden Empire doesn't need to use any shell companies to make their system work. They already have powerful people in the right positions in all the global monetary institutions to make everything work the way they need it too."

Todd narrowed his eyes. "And we know they'll eliminate anyone who becomes a serious threat to interfere with it working smoothly. The team of financial experts we're working with can't untangle the mass of information we've collected fast enough to make any quick progress."

Judy nodded. "We estimate it will take over a hundred years just to untangle what we have now. It's no wonder the Hidden Empire has been able to remain basically invisible for a long period of time."

Ralph leaned back in his chair. "Well, you need to go back to spending most of your time on our own finances anyway. We need to step up the farming activities on all the new land we've acquired for growing poppies, and there a lot of financial work involved."

"We'll take care of that," Judy said. "But Todd and I each have a strong desire to unravel the global money flow of the Hidden Empire and understand it. We'll take care of our needs first, but I think it's in our best interest to keep working on the

Hidden Empire finances too."

Ralph grunted. "Okay, but plan to spend most of your time on Tuxtun's business finances."

Judy and Todd both nodded.

Matt glanced at Ralph. "Before we end the meeting, I would like to get an update from Todd on our latest technology issues, and an update from Steve on our security situation."

Ralph looked at Ben.

"I think hearing these updates will be good for me too," Ben said. "That will help me in my planning to give you the necessary support. Tomorrow, I'm getting an update from Kyle on our technology situation at the FBI." He glanced at Ralph. "My getting these updates over here from your team will help me in several ways."

Ralph nodded and looked at Steve. "You go first."

"Since you're having the meeting with Belzex here in the next few days we'll beef up security even more," Steve said. "You're already having the Belzex group check their weapons before they come inside the building, but I'll still position a lot of security just outside the conference room."

"That's good," Ralph said. "We need to be prepared for anything. We'll have our pistols, but no telling what kind of tricks they might pull."

Steve explained some more security details and then gestured to Todd.

"We've studied a lot of things together and I think we both agree that we've covered everything as well as we can," Todd said. "We've tried to leave nothing uncovered." He went on to give a ten-minute summary of their discussion on the overall security measures."

"So, you're even going to have reinforcements ready in the hallway if they bring any of their big Chinese assassins with them." Matt grunted. "That's a good idea."

"Like Ralph mentioned, we want to be prepared for anything," Steve said. He glanced at Ralph and Matt. "From your descriptions of your previous encounters with the big

Chinese guys, they can't be stopped immediately, so I've arranged for some of our biggest and toughest security guards to be available to assist."

"I'm glad you thought of that," Ralph said. "Since they can't bring weapons into the building, they'll likely have some excuse for bringing one of those big guys."

Matt shrugged. "We could just not allow him to come into the building."

The discussion continued for another hour, and many topics were discussed before everyone concluded they were as prepared as they could be.

Chapter 20

In Mexico City, Leon gathered all the Belzex leaders together in a large lounge at the mansion of Jorge Soto to discuss their strategy in the meeting with Tuxtun, which was now scheduled for an afternoon the next week.

Ivan looked at Leon. “Since we can’t take our guns inside the building, do you still have a strategy for the upcoming meeting with Tuxtun?”

“It’s basically the same as before,” Leon replied. “We’ll be congenial and make an offer. If they don’t accept it we’ll negotiate a bit. If they refuse to sell at any cost, one of our assassins will kill Ralph Gibson and Matt Baker.”

“Since we won’t have any guns, how can it be the same as before?” Chen asked.

Everyone looked at Leon.

Leon turned toward Chen. “Have you recovered our application that will make someone invisible to all security monitors yet?”

“I’ll check on the current status,” Chen said. “We have the backups to all the old Belzex applications setup in a warehouse we bought. We’ve been activating them and testing them as fast as we can. I know that we have some of them working."

Chen removed his phone from a pocket and pushed a button. After a short conversation, he looked at Leon. "It's been activated and tested." He flashed a big smile. "It's ready to use. While we're waiting for our new hi-tech fortress to be built we have all our technology applications assembled in a warehouse undergoing testing."

Leon gave a thumbs-up.

"We'll put it to good use and this time we'll get Baker and Gibson out of the picture for good," Leon said. "We'll leave the door wide open for whoever takes over Tuxtun to sell us the land."

"What's your plan?" Ivan asked.

"Since we can't take any weapons into the building, we'll make sure our weapons are already in there waiting for us," Leon said. "And then the plan will be very similar to what we did last time."

Everyone stared at Leon and waited for him to continue.

Leon glanced at Chen. "You pick the person or team who goes, but we'll plant the weapons inside the Tuxtun headquarters building the night before we have the meeting. We'll test the invisibility app again to make sure it's working properly and then we'll use it to plant the weapons."

Chen nodded and waited for Leon to continue.

"We'll take two assassins to the meeting with us. We won't use Rafa since Baker has already had an encounter with him." Leon glanced at the faces around him. "We'll use the two who're the closest to looking like businessmen."

"What weapons and how many are you going to plant inside the building?" Jorge asked.

"Two pistols," Leon said. "We'll use one against Baker and one against Gibson. Since they're both good with pistols, and I'm sure they'll be armed, we'll must fire both our pistols at the same time."

There was a brief silence.

"Where will the pistols be stored?" Ivan asked.

"The meeting was held in their main conference room, which they call their boardroom, last time," Leon said. "I'm guessing they'll hold it there again. We'll tape the pistols under the top of the table."

Ivan leaned forward. "How will you know where to tape them?"

"In the last meeting, all the Tuxtun group sat on the opposite side of the table from the door," Leon said. "I'm guessing that's where they'll sit again. It's logical for them to want to watch for anyone entering the room unexpectedly. It's a good safety precaution."

Everyone nodded slowly.

"We'll make sure our assassins sit in the two seats at the corner by the door on our side of the table," Leon said. "I think we can do that without arousing any suspicion in any way."

No one commented immediately. Everyone seemed to be thinking.

"The assassins we take with us will need to practice removing the pistols from the tape and from under the table," Chen said. "They must be sure to be inconspicuous in all their movements." He paused. "And they'll need to practice getting their pistols into position to fire quickly and accurately."

Leon nodded. "Of course." He gestured to Chen. "You'll be in charge of selecting the best pistol to use and for any training that you think is necessary."

Chen gave a thumbs-up. "There are a lot of specialty pistols we could get, but I think the standard .44 Magnum revolver will suit our purpose well."

Everyone seemed to agree.

After a brief pause, Bolivar asked, "So do we have the same strategy as before that after our assassins shoot them, they'll immediately drop their pistols and each raise both hands in the air?"

"Yes, of course," Leon said. "The rest of us will act shocked they would do such a thing. We'll call the police immediately and they'll be arrested."

"But we'll assure them beforehand we'll do whatever is necessary to get them out of jail quickly, right? Jorge asked as he stared at Leon.

"We'll have no problem doing that," Leon said. "Much like the Hidden Empire, we have both the money and the means to influence the right people to make things go the way we need them to go."

A brief silence ensued.

"We'll still need to be careful with Baker and Gibson. They'll be armed and we know from experience that they respond quickly to danger." Leon looked at Chen. "You'll need to keep that in mind during your training."

Chen nodded.

"It will take each assassin at least a second to move his pistol above the table and shoot,". Bolivar said. He scanned the faces around him.

"I'll take that into consideration," Chen said. He stared at Bolivar for a moment. "You can come watch the training if you would like."

"I'm sure you'll cover everything okay." Bolivar gave a wave of dismissal.

A brief silence ensued.

Jorge looked at Leon and Chen. "Other than the assassins, none of us will have any weapons. What if someone at Tuxtun besides Baker and Gibson shoots back? They have a security chief and he's probably a good shot. We expect them all to be carrying weapons in this meeting." He paused. "Do we have any plans for that?"

Everyone stared at Leon. Obviously waiting for him to say more.

Leon hesitated for a moment and then looked at Chen. "We could take one of your monster Chinese assassins. What do you think about that?"

Chen shrugged. "If they won't let him inside the building, we have nothing to lose by trying."

"Yeah, you're right. We'll be smart to give it a try" Leon

said. He paused for a few seconds. "He would give us an extra option to reach our objective if the first part doesn't go as planned."

Chen narrowed his eyes. "He can strangle people quickly when necessary and he can get to them fast."

"We'll be offering a huge amount of money for the land. If things go well, we might not need to kill anyone," Leon said. "I just want to be prepared in case we do."

Everyone nodded.

There was another brief silence.

"There's something that's bothering me," Chen said. He frowned and looked at Leon.

Leon hesitated for a moment and then gestured to Chen. "What?"

"We know the people at Tuxtun are smart. I'm guessing they'll consider the possibility that we can use the technology we used before to break into their headquarters without being detected," Chen said. "They'll scan the meeting room with a metal detector for weapons and definitely check under the table if their scan doesn't detect any."

"So, using some of our special pistols that can't be detected by a typical scanner won't work?" Leon asked.

"That's right, pistols made from our special material constructed from a mixture of polymer and nylon can't be detected by a typical scanner, but they're visible to the eye," Chen said. "I think the Tuxtun group will be smart enough to look under the table and the chairs as part of their security check before everyone comes into the conference room for the meeting."

Leon nodded slowly and then scanned the faces around him. "Okay, any suggestions?"

Ivan looked at Chen. "We'll still have one of your big Chinese assassins as backup, right?"

Chen nodded. "I think Yang Chee is the one I'll bring. He and Huang Ban are both good, but I think Yang is the faster of the two, and that could make a difference."

Ivan narrowed his eyes. “Okay, so if we can get him in the meeting as our bodyguard, we’ll have our backup in place.” He paused and looked at Leon. “I think we’ll use two pistols that won’t set off metal detectors and we’ll store them in a bathroom that’s close to the conference room.” He kept a steady gaze focused on Leon.

“Yeah, I noted last time that they had one in the hallway by the conference room,” Leon said. “They must have one available for people who attend the meetings.”

Ivan rubbed his chin. “We’ll use a similar plan to what we discussed before, but we’ll put them in the bathroom instead of the conference room. The person who breaks into the building can look around in the bathroom and decide where to put them. He should be prepared to tape them inside a cabinet where they won’t be spotted easily if the cabinet doors are opened.”

The discussion continued for a few more minutes and then Leon scanned the faces around him. “This is what you call careful planning.” He leaned forward. “We’re preparing for all contingencies. We should be successful.” He grinned and gave a thumbs-up.

Everyone else gave a thumbs-up too.

Chapter 21

At Tuxtun, Matt and Ralph sat at the round conference table in Ralph's office with seating for eight.

"I've been informed seven Belzex executives plus a bodyguard are planning to attend the meeting we're having with them tomorrow about selling the land," Ralph said. "I wanted to talk that over with you before we discuss this with Judy, Todd, and Steve."

"Are you planning to invite Ben?"

Ralph shrugged. "Probably. What do you think?"

"I think it would be a good idea. He would be an extra layer of protection if they tried to pull off a stunt like they did last time."

"Remember, they're having to check their weapons before they come into the building this time."

Matt nodded. "But they might have some other tricks up their sleeves."

"Such as?"

"A lot of things come to mind."

Ralph checked his watch and regarded Matt carefully. "We have plenty of time. Give me some examples of the things you're thinking about."

"Remember how they had the technology to unlock our doors and make themselves invisible to all of our security systems?"

Ralph hesitated for a moment and then leaned forward. "They don't have any of that now. The Hidden Empire destroyed everything when they put them out of business before."

"We can't be sure about that."

There was a brief silence.

After a long pause, Ralph nodded. "You have a good point there."

"We know, for some reason, the Hidden Empire allowed Leon Schulz to live. We can't be sure they took away all the technology Belzex was using."

Ralph narrowed his eyes and took a deep breath. After he exhaled slowly, he looked directly at Matt. "Tell me more about some of the other things that came to mind."

"If they have their technology back, they can get into our building at any time." Matt paused and tightened his lips. "They could come in at night and deposit weapons in some hidden places inside the building."

"Do you have any guesses as to where those places might be?"

"Somewhere along the path from the front lobby to the conference room, or even in the conference room."

Ralph grunted. "Keep thinking. We need to outsmart them."

"I would tape the weapons under the conference room table on their side of the table."

"How will they know which side of the table they'll be on?"

"Leon is pretty smart. During our last meeting we sat on the side of the table that allowed us to face the conference room door. We do that for all our meetings here at Tuxtun for obvious security reasons. Leon probably took note of that during our last meeting here and figures we always do that."

Ralph nodded slowly. "So, they'll expect to be seated on the side of the table where their backs will be toward the door."

"Right," Matt said with a grin. "If they have their advanced technology back, I think we can figure out what they're going to do."

Ralph nodded slowly and then regarded Matt carefully. "So, I guess we'll sweep the room for weapons with our detection equipment before the meeting and do a visual inspection too?"

Matt narrowed his eyes. "Yep, I think you have that exactly right."

Ralph set his jaw and stared at Matt. "Before we call in Judy, Todd, and Steve, should we discuss some more things?"

Matt gave a slight shrug. "Maybe a couple of more things, and then we'll go back over all this again and get their thinking on it."

"Okay, what's next?"

"I'm guessing the bodyguard they want to bring with them will be one of those giant Chinese guys," Matt said. "They'll figure that will give them another option to eliminate us in case their other plans don't work out."

Ralph nodded. "We'll be armed and we recently went to the FBI training facility here and completed another tune-up session with our pistols. We'll plan to be as well prepared as possible. We need to make damn sure we don't get caught by surprise. We'll for sure stay on our toes."

"It will be interesting to see who they bring to the meeting," Matt said.

"Yeah, they might even bring some additional people. We didn't give them any limit."

"Matt looked at Ralph and grinned. "We're pretty good at sizing up assassins. If they do bring an extra person or two, we'll assume the worst."

There was a brief silence.

Matt leaned forward. "Getting back to business, do you

have any specific objective for this next meeting?"

Ralph grinned. "I think you nailed it earlier. We'll hear what they have to say and tell them we'll think about it and get back to them."

"I don't think they'll want to wait. We should expect them to insist on making a deal before they leave."

There was a brief silence.

"Anything else you want to discuss?" Ralph asked.

Matt thought for a few seconds and then shook his head slowly.

Ralph glanced at his watch. "Guess it's about time to bring in Judy, Todd, and Steve and get their opinions about our plan for the meeting."

Matt nodded. "I'm glad they've had plenty of training and can defend themselves if they need to." He paused for a moment. " And I think we're all experienced enough to adjust appropriately if anything surprises us."

Ralph punched a button on his phone. "We'll meet in the boardroom." He and Matt both stood.

Chapter 22

A few minutes later, Judy, Todd, and Steve entered the boardroom and sat across from Ralph and Matt at the large table that could accommodate twenty people, seven on each side and three at each end.

Ralph and Matt took about ten minutes to explain their thinking about the meeting with Belzex.

"Any guess about whether they might have the technology working again?" Judy asked.

Ralph shrugged. "I just know they might have it working again and we need to assume it is and check everything." He glanced at Steve.

"I'll have my security team do a thorough check just before the meeting," Steve said as he glanced at Ralph. "I know it's tomorrow, but I don't know the time."

"Two o'clock in the afternoon," Ralph said. "You decide when you'll do your checking."

Steve nodded and made no further comment.

"Do we want to keep our pistols hidden or let it be obvious we have them?" Todd asked.

"Do you have a reason for asking that question?" Matt

asked.

"Just wondering if we would have any benefit in keeping them guessing," Todd said. "I thought it would be good to know exactly what our strategy is so we could all work together to execute it."

"We'll keep that question in mind and discuss it tomorrow," Ralph said. "The meeting is at 2 PM so we should have plenty of time in the morning to decide what's our best strategy is."

Everyone nodded.

Chapter 23

In Rome, Italy, four elderly men sat in cushioned chairs inside an ancient but elaborately furnished mansion.

Vulcan and Neptune each filled his pipe with tobacco. Zeus didn't smoke and he knew Apollo didn't smoke but, as usual, he patiently waited for Vulcan and Neptune to get ready to give their full attention to what he was going to tell them.

Vulcan and Neptune each lit his pipe and took a few puffs and then they glanced toward Zeus.

Zeus always wanted their full attention, so he always exercised a lot of patience in waiting for them to get ready to fully absorb what he was going to tell them.

In a few more minutes, Zeus said. "We had a good discussion about our history in our last meeting, and this time I want to concentrate on fully understanding our strategy in using the Specter Matrix."

Everyone nodded.

"Do we have some unexpected problems?" Apollo asked.

"We don't have any problems yet in the actual execution, but some new things have come up that need to be addressed," Zeus said. He leaned forward. "During our extensive analysis of all possible threats we've uncovered some potential issues that need to be resolved. I always want to take every opportunity to ensure that all our members are as knowledgeable about all the potential threats concerning us and our organization as possible."

Everyone waited for Zeus to continue.

"I've told you before Jupiter, Odin, and I think it's good to make sure everyone in our organization knows our complete history and I've mentioned many times you can't absorb all of it in one session," Zeus said. "So, we'll cover a little more history, and I'll bring you up to date on the Specter Matrix activity."

No one commented.

"We've discovered the NSA is using a lot of stochastic processes to try to track our cash flow through our global monetary system." Zeus leaned forward. "A stochastic process is a mathematical model used to describe systems or phenomena that evolve over time in a random or unpredictable manner."

"Why is it important for us to know this?' Vulcan asked.

"Yeah," Neptune said. "We have experts who take care of these details so we can concentrate on where the money needs to go."

Apollo looked at Vulcan and Neptune but didn't make any immediate comments.

Zeus leaned forward. "You're responsible for making a lot of the important political decisions, but I've found over the years that's it's also important to have a grasp of the technology aspects involved in carrying out the decisions you make."

Vulcan, Neptune, and Apollo all remained silent and stared at Zeus, obviously waiting for him to say more.

"If we have some problems with the cashflow and we need some explanations from our experts, it's good to have some idea of what they're talking about," Zeus said. "If someone you're trying to transfer a large amount of money to doesn't get the money, it's good for you to have some idea of what the problem is and what kind of explanation you might want to give to the person who's wondering what happened. That can be really important at crucial times"

Everyone nodded.

"I completely agree," Apollo said. "If you know a little more of the technical details, I'd like to hear them."

Zeus looked at Vulcan and Neptune.

They both shrugged.

"Apollo is a good bit younger than we are," Neptune said. "He has more of an appetite for these details, but we can handle it. Go ahead."

Vulcan nodded.

"Okay, here we go." Zeus grinned. "I like this stuff, but I'll try not to overdo it." He paused. "Formally a stochastic process is defined as a collection or family of random variables indexed by a specific parameter, which typically represents time."

Zeus looked directly at Vulcan and Neptune. "It's important to note that random variables and time were involved."

They both nodded.

"However, speaking of time, we do have the Markov Property to consider, which at first glance is a contradiction," Zeus said. "It says the probability of future states in a stochastic process depends solely on the current state, completely independent of past history. In essence, the present summarizes all necessary information about the past."

Vulcan shook his head slowly. "I find that confusing."

Neptune looked at Zeus and nodded.

Zeus explained.

"It's still a bit confusing," Apollo said, "but I get the drift."

"The main thing to remember is that our experts are dealing with all that and many other technical aspects in trying to keep the money flow in the Specter Matrix undetectable." Zeus paused. "It's good for all of us to know our experts are constantly involved in that type of activity so we can remain titans of influence across the globe. The necessary amount of money must get to the right individuals and organizations without anyone else knowing about it."

Everyone nodded.

There was a brief silence.

Zeus rubbed his chin. "Well, before we proceed to some other subjects we need to cover, does anyone have any questions?"

"We still consider ourselves the Illuminati, but everyone in law enforcement refers to us as the Hidden Empire." Vulcan took a long puff on his pipe and stared at Zeus. "Any change in that?"

Zeus shrugged. "Not really, but we sometimes use the term Titans of Influence to describe us. To my knowledge, none of us use the term Hidden Empire."

Vulcan nodded slowly and didn't comment.

"You should always feel free to question what we're doing and how we're doing it." Zeus paused. "And remember, we'll always go to great lengths to ensure all our locations are constantly protected from eavesdropping, no matter how good the spying technology is by anyone else."

There was a brief silence.

Apollo looked directly at Zeus. "The Knights Templar organization contained the most famous heroes of the Crusades and when the Pope abolished them, they joined the Free Masons. As things evolved the inner circle of the new combined group formed our organization and we became known as the Illuminati." He paused a moment and narrowed his eyes. "Another name for the Illuminati is The Enlightened Ones. Do you know what that's based on?"

"Illuminati is a Latin word," Zeus said. "The Enlightened Ones is just the translation."

"And do you and our other leaders still want a one-world government and a religion that controls the population?" Apollo asked.

"Yes, that's the long-range goal, but, as I've mentioned before, we're not rushing toward that," Zeus said. "We eventually plan to abolish private property, but we've decided to move forward very slowly on that. Some prestigious global groups are largely controlling the pace."

Apollo leaned toward Zeus. "Yeah, I remember you telling us that some of these groups are active in forming many aspects of our long-range plans and goals."

Zeus nodded. "Odin, Jupiter, and I are active members of most of these groups. Some of these groups have meetings around the world that are closed to the public. A couple of the groups have an annual, unofficial, invitation-only conference of people from North America and Western Europe." Zeus smiled and waited for some questions.

"And you told us about a third of the people attending these meetings are from government and the rest are from finance, industry, and education," Neptune said.

"That's correct." Zeus leaned forward. "At many of these conferences we discuss major problems facing the United States and Western Europe. Academia is often a big player where representatives from a lot of prestigious universities are involved."

"You come up with action plans at these conferences and that's how you control the pace of the changes that are made, right?" Vulcan asked.

"Exactly," Zeus said.

"When are you going to discuss the latest action plans?" Apollo asked.

Zeus flashed a slight grin. "We've actually been doing that. I've mentioned before that certain things sometimes get very complicated, so everything is not always immediately clear. Sometimes you need to have patience."

Everyone nodded.

"The tasks you've been given over the last few years are examples of various aspects of our action plans," Zeus said. "And remember we have to modify our action plans at times with certain specific actions."

Vulcan narrowed his eyes. "Yeah, you've reminded us frequently that you must take drastic actions sometimes for the protection of our organization. The situation with Hugo Wagner was one of them."

"Hugo was useful for a while. The disruption he caused made it easier for us to keep all our dealings hidden from public view," Zeus said, "but he crossed the line when he made plans to get all the gold in the United States, shoot missiles into the United States, and attempt to conquer the world." He paused. "Hugo got some stern warnings for all of that, but he ignored them." Zeus scanned the faces around him. "Getting all the gold would have disrupted our ability to maintain our flow of money which we need to continue to operate. And it should be obvious why we wouldn't want him shooting missiles into the United States."

Everyone again nodded and made no comments.

"We must constantly payoff all the right people," Zeus said. And we must pay some of them a huge amount of money to steer some important world events in the direction we need them to go." He regarded the people around him carefully. "The Russian Government is indirectly involved in what we do because we control the Russian KGB and when we need them, they provide hit squads and heavy weapons, like the missiles we used on Hugo."

"When you need the KGB to take action, do you need to pay off the Russian leaders?" Apollo asked.

"That's usually the case," Zeus said. "I'll make sure all of you have more insight into that as things move along with Leon Schulz."

Vulcan, Neptune, and Apollo all nodded slowly.

"And remember our power stretches a long way back. Remember we had to destroy the Titanic back in 1912 to meet some important goals," Zeus said.

Vulcan, Neptune, and Apollo all nodded.

There was a brief silence.

Zeus scanned the faces around him. "It looks like the three of you have retained well the information we've discussed in earlier meetings, so I'm not concerned about needing to reinforce the information we've discussed previously. We'll move on to some new things I need to mention to you." He paused. "Of course, you wouldn't be part of this organization if you didn't have exceptional capability to help execute our plans."

Zeus explained some more and then asked, "Well, does anyone have more comments or questions before we continue?"

"In every meeting you've conducted with us over the past few years, you've always emphasized maintaining the money flow we need, right?" Apollo asked.

"Right, that's the primary focus of our operation. Without the constant money flow, we can't function like we need to," Zeus said. "And constantly keep in mind we've had to do a lot of drastic things to maintain that money flow."

Everyone nodded.

"Leon Schulz became a big problem before. How do you feel right now about him becoming a big problem?" Neptune asked.

"Like I've mentioned, we'll constantly keep a close eye on him and Belzex," Zeus said. "Just like before with him and Hugo, the disruption is useful, but if he gets out of his lane we'll terminate his operation like we did before , and this time we'll also eliminate him if we need to do so."

Vulcan leaned toward Zeus. "Thanks again for keeping us well informed."

"You're welcome," Zeus said. He scanned the faces around him. "Any more comments or questions before we move on?"

Everyone remained silent.

"Now we'll discuss a few more details you need to know and then we'll wrap up," Zeus said. "It's good you retain the information as well as you do."

No one commented. They obviously were waiting for Zeus to continue.

"In keeping with ensuring you're well informed, I want to make another point about the Specter Matrix," Zeus said. "Our experts use a transformation matrix as part of the functionality of The Specter Matix. That's a matrix that transforms one vector into another vector by the process of matrix multiplication. The transformation matrix alters the cartesian system and maps the coordinates of the vector to the new coordinates."

A brief silence ensued.

"Are you going to give us a few more details?" Vulcan asked.

"I didn't think you and Neptune were interested in more details," Zeus said.

Vulcan gestured toward Apollo. "We know he is and we can tolerate it" He looked at Zeus. "Besides, if we get involved with anyone who didn't get the money they were expecting, it will help to have a little more knowledge about the subject matter when we talk to our experts."

Zeus grinned and nodded slowly. "Okay here we go. Remember our experts are constantly updating the Specter Matrix to make sure it continues to function properly in spite of everchanging technology, and ensuring it functions properly involves a lot of things, including advanced mathematics."

Everyone nodded.

"The Cartesian coordinate system is a grid-based mathematical map used to uniquely identify the location of points in a space using numerical coordinates. It bridges algebra and geometry, allowing equations to be visualized as shapes and graphs," Zeus said. "The combination of origin and basis forms something called the Cartesian frame, and the position of any point in three-dimensional space can be specified by three Cartesian coordinates, which are the signed distances from the point to three mutually perpendicular planes. These coordinates specify the point in an *n*-dimensional Euclidean space."

Neptune shook his head slowly. "That's a little much for me right now." He paused and looked at Zeus. "How do you retain this level of technical knowledge?"

"I have a PhD in physics," Zeus said. "I understand all this stuff and that makes it easy for me to talk to our experts who maintain the Specter Matrix."

"Do What about Odin and Jupiter, do they have similar degrees?" Vulcan asked.

"They each have an advanced degree in finance," Zeus said. "We make a good team." He grinned.

"Would you elaborate some about the Euclidean space?" Apollo asked.

"Euclidean space is the fundamental, flat space of classical geometry. It represents physical space, in our everyday experience," Zeus said. "It generalizes the familiar 2D plane and 3D space to any finite number of dimensions. It allows for standard distance measurements using the Pythagorean theorem, and lines that are parallel remain equidistant infinitely. It helps in defining the pathway throughout our multidimensional Specter Matrix."

Neptune continued to shake his head. This time Vulcan joined him.

Apollo gave a thumbs-up.

"To model how money changes hands over time, economists transform transaction tables into stochastic matrices," Zeus said. "This allows the circulation of money to be viewed as a Markov Chain Process."

Apollo rubbed his chin and leaned toward Zeus. "And a Markov Chain Process is?"

"It's a mathematical system that models a sequence of random events. It uses the Markov property, which relates to the probability of transitioning to the next state depends only on the current state and not on the sequence of events that preceded it." Zeus scanned the faces around him. "Basically, you need to remember that the logic driving the Specter Matrix involves a lot of advanced mathematics and the latest innovations in quantum computing." He paused. "If you remember just a few of the specifics, that could be useful at certain times."

Vulcan, Neptune, and Apollo all nodded.

There was a brief silence.

" Would you tell us about some of the latest innovations in quantum computing?" Apollo asked.

Zeus gave a slight shrug. "There are a lot of them. I don't think you want all those details."

"Would you give us a few?" Apollo asked.

Zeus looked at Vulcan and Neptune.

"It would be good to know a few highlights," Vulcan said. He glanced at Neptune.

Neptune gave a thumbs-up.

"We use a fully operational Quantum Financial System (QFS) in the Specter Matrix," Zeus said. "That makes full use of quantum algorithms for risk analysis and quantum-secured networks."

Everyone nodded but made no comments. They were obviously waiting for Zeus to say more.

"Quantum finance is an active research activity targeted toward improving cybersecurity," Zeus said. "Our experts are actively involved in that. It features the use of light particles (photons) to transmit financial keys. Anyone observing the financial state resulting from this activity alters it and sets off an alarm."

The discussion continued for a few more minutes and then Zeus stood.

"Remember, we must be able to control important events in the world. Jupiter, Odin, and I always try to keep our members fully informed and fully aware of the reason drastic measures need to be taken at times," Zeus said. "To continue our power, we must have the means to remain masters of stealth and titans of influence. We must always do what is necessary to maintain our money flow."

A brief silence ensued.

"Now you're up to date and you have some additional details you need to keep in mind as we move forward," Zeus said.

Vulcan, Neptune, and Apollo all gave a thumbs-up.

Chapter 24

At Tuxtun the next morning, Ralph sat in the boardroom along with Matt, Judy, Todd, and Steve.

Ben and Kyle from the FBI were also there.

"We'll take a little more time to go over our strategy before the meeting with Belzex," Ralph said. He scanned the faces around the table and then focused directly on Ben and Kyle. "We've worked with both of you for quite a while now. Don't hesitate during our discussions to bring up any concerns you have."

Ben grunted and leaned toward Ralph. "You don't need to ever have any concern about that, and I have a question to start with."

Ralph gestured for Ben to go ahead.

"You already know what to expect from Leon Schulz and I don't think you're going to sell the land at any price, so why are you even having this meeting?" Ben asked.

"You told us before that the only member of Belzex who survived the onslaught of the Hidden Empire was Leon Schulz," Ralph said.

Ben nodded. "That was all confirmed by the CIA and

several other intelligence agencies."

"Well, he has a whole new team now, and I want to take a look at who we're going to be dealing with this time," Ralph said. "Knowing a little bit more about them could turn out to be valuable as we go forward in having to protect ourselves from the things they'll probably do."

"Makes sense." Ben gave a slight laugh. "So, you're taking advantage of a good chance to size them up."

Ralph grinned. "Yeah, and we'll each be wearing a Glock 19 in a holster positioned toward the back of the right hip. We've been doing that in the office ever since you recommended it a year ago."

"Our pistols are always hidden," Judy said. "I'm in the habit of wearing slacks and a sports coat every day. That works well for me."

Ralph leaned toward Ben. "And thanks to you, we've had a lot of great practice in using the pistols."

Ben gave a quick nod. "Let's discuss the strategy you're going to use this afternoon."

"It keeps changing," Ralph said. "It's good to all be together to try to make sure we're not overlooking anything." He gestured toward Steve.

"We figured they were going to be hard to outsmart and, so far, they're proving that to be true." Steve scanned the faces around the table. "Somehow they have use of at least some of their previous technology and they broke into our building last night."

"Using the technology to make themselves invisible to all of your security systems?" Kyle asked.

Steve nodded and gestured toward Todd. "He helped us figure out the things they might do, and, with all the possibilities in mind, we checked a lot of areas inside and outside the building."

Todd gave a thumbs-up.

"He and my security team did a thorough job of brainstorming, and we discovered they planted two pistols inside the bathroom across the hall from the boardroom," Steve said. "The pistols were undetectable using metal detectors because they were made from a polymer/nylon mix and the bullets were made from the same stuff, but we found them with a thorough visual search."

Todd flashed a big grin. "They taped them inside a cabinet and at the top of the cabinet. They did a good job because they weren't visible when we opened the cabinet doors, but we thought about what we would do if we wanted to hide pistols in there, and we found them."

"I'm assuming you did the same search in all the various areas, especially here in the boardroom." Ben leaned slightly toward Steve.

"You bet," Steve said. "At first, of course, we figured they would tape the pistols under the top of the table in here, but when we didn't find them we checked all the other possibilities."

Todd nodded. "That's when all the brainstorming came in handy. By the way, we're all good at it. It's good to stay sharp on a lot of methods and techniques in this modern era."

The discussion continued on a lot of subjects relevant to the meeting until they took a break for lunch. One of the things discussed was whether or not to let one of the giant Chinese into the meeting if he came with the group from Belzex.

*****

After lunch, the group at Tuxtun reassembled in the boardroom a little before 2 PM.

"In a few more minutes, Ralph's secretary escorted eight people into the boardroom. All the Tuxtun group stood to greet them.

Introductions were made all around. The eighth visitor was indeed a giant Chinaman introduced as Yang Chee, the bodyguard.

Leon Schulz continued to introduce Chen Wan, Ivan Rowan, Bolivar Ortiz, Jorge Soto, Stepan Petrov, and Wolfgang Fischer.

When everyone took their seats, Matt noted Stepan and Wolfgang sat at the end of the table by the door, which was expected. As per the Tuxtun plan, all the Belzex group had their backs to the boardroom door. All the Tuxtun group were seated facing the door.

Yang Chee stood in the far corner on the Belzex side away from the door.

Matt figured Yang was too heavy to sit in a regular chair and they wanted him to stand anyway.

Stepan Petrov and Wolfgang Fischer had been introduced by Leon as investment bankers who controlled large financial institutions. However, Matt knew better. He knew they were assassins and they were soon going to get a big surprise when one of them checked the bathroom.

Ralph looked at Leon, who was sitting directly across from him, and said, "So you're again interested in buying some of our land?"

"Our business is booming. The demand far exceeds the supply," Leon said. "As before, we want to expand our Poppy crops, and you own all the prime land around our existing crops."

Ralph leaned forward. "And, of course, you only want to farm on prime land."

"There's no need to make any secret of that. That's always been our objective. We want to produce the best Morphine and Codeine products possible." Leon regarded Ralph carefully. "We do that now and we want to keep it up."

"Producing the best Morphine and Codeine products possible?" Matt asked.

"Yes," Leon said. "And that's a very reasonable objective." He extended both arms to the side with his palms up.

"And you mentioned the demand for your products far exceeds the supply," Matt said. "So, your interest in acquiring more land is logical."

"We always want to produce the best quality products and we're willing to spend a lot of money to get what we need." Leon looked back at Ralph. "So, are you willing to consider a really good offer?"

"It won't hurt anything to listen to your offer," Ralph said. He leaned toward Leon. "Let's hear it."

"We offered you ten million dollars an acre last time," Leon said.

Ralph drummed his fingers on the table for a moment. "I remember that well."

"And you didn't take it," Leon said.

"I also remember that well." Ralph continued to drum his fingers on the table. "And I don't plan to take it now. We'll do better in the long run by keeping our land."

Stepan slowly rose from his chair. "I think I noticed a bathroom across the hall from us. If you would excuse me for just a minute."

No one commented as Stepan opened the door and moved into the hallway, closing the door behind him.

Matt figured things were going according to plan for them so far now that it was beginning to look obvious that Ralph wasn't going to sell any land. It was likely Stepan was going to get the pistols so that he and Wolfgang could prepare to shoot the two top executives in Tuxtun, Ralph and him.

Leon leaned toward Ralph. "As soon as Stepan comes back, I'll find out if he's received any of the big deposits he was expecting from some of our biggest clients."

Matt knew what was going on and he was certain everyone else in the Tuxtun group knew too. Stepan was about to get a big surprise. It would be interesting to see how he handled it.

"Are you planning to increase your offer for our land?" Ralph asked.

Leon nodded. "If we have the extra money."

Ralph stared at Leon and his gaze became more intense. "How much are you thinking?"

"We might be able to go to twelve million dollars per acre," Leon said. He looked at Wolfgang. "Do you know if Stepan has received any of those big deposits we've been expecting?"

Wolfgang shrugged. "We have totally separate operations, and we both stay very busy. We don't have much time to interact very often."

Leon nodded slowly just as the door opened.

Stepan walked into the boardroom, closed the door behind him, and went straight to his chair.

Matt thought that Stepan handled things in a very professional manner. He showed no signs of panic or disappointment.

In a couple more seconds after Stepan had settled into his chair, Leon looked at him and asked, "Have you received any of those big deposits you were expecting?"

"No," Stepan said. "I'm disappointed. Those two big ones I was expecting didn't arrive." He glanced at Wolfgang. "We're going to need to alter our plans for the moment."

Wolfgang tightened his lips and nodded.

"I thought you guys didn't interact," Matt said.

"We had a special plan I told them we were going to discuss if the two big deposits had arrived," Leon said. He gestured to the Belzex group on his side of the table. "Now we're going to need to use our alternate plan."

Matt figured the big Chinese guy would be going into action at any moment. He was ready and he knew all in the Tuxtun group were ready. They had discussed this probability at length this morning.

Leon looked at Ralph. "Well, I can't make a better offer right now, but I can still offer the ten million per acre if you want to reconsider."

"I asked you about this before, but have you given any additional thought to buying some prime land in other areas?" Ralph asked.

"Of course," Leon said. "I constantly keep it in mind, but the land we own is the best in the world for growing poppies.

As I mentioned, we want to expand our production, and you own all the rest of the prime area around our property." He leaned forward. "We want only the best land. Our quality will suffer if we grow our poppies on inferior soil."

Ralph nodded slowly and didn't comment.

Leon leaned toward Ralph. "We're making you a great offer for your land. We know that's a lot more than you paid for it."

Ralph shrugged. "Like I told you, we're not interested in selling. We want the best quality too."

Leon narrowed his eyes. "Eight billion dollars is more than your profit in a whole year. I checked that yesterday to make sure that's an accurate statement."

"If we keep the land, our profits will improve rapidly once we start selling the additional products from the poppies," Ralph said. "We're seeing the increasing demand for Morphine and Codeine too."

Matt knew that Leon was planning on producing a lot more heroin and opium in addition to the legal drugs, and that would bring him huge profits.

Yang took a step toward the side of the table where all the Tuxtun group sat and glanced at Leon.

Matt and Ralph stood and drew their pistols within about a half second. Everyone else in the Tuxtun group took slightly longer, but, within two seconds, Yang had seven pistols pointed at him.

Leon motioned to Yang to stop and then looked back at Ralph. "We've made you a very generous offer. Do you want to think about it for a while?"

"That's not necessary. We've already thought about it," Ralph said. "We're not selling at any price."

Matt could tell Leon was considering turning Yang loose, but he apparently decided not to. Matt figured Leon had determined there was no way Yang could live long enough to strangle both their targets.

"There's no need to stay any longer then," Leon said. He

stood and quickly waved all his team toward the closed door of the boardroom.

Stepan got there first and opened the door.

As the Belzex group started moving out into the hallway, Matt saw that Steve's security team was out there to escort the Belzex group to the building exit.

"Sorry to waste your time," Leon said over his shoulder, "let me know if you change your mind."

No one in the Tuxtun group commented. They all kept their pistols extended toward the Belzex group until Steve's security team had escorted them out of the area.

Ralph glanced at his watch and then said, "We'll take about a thirty-minute break and then reconvene in here to discuss things and decide what our priorities are moving forward."

*****

Thirty minutes later, the group of seven reconvened their discussion.

Ben looked at Ralph and Matt. "So, they're for sure wanting to get you out of their way."

"Yes," Matt said. "They've obviously concluded they must have the land to meet their goals, and they had decided beforehand to eliminate Ralph and me if we wouldn't sell."

Ralph nodded. "My thoughts exactly."

"Over the years I've learned a lot from dealing with their types," Ben said. "All of them are extremely greedy. They have an overwhelming urge to make as much money as they can in every way they can. Drug trafficking is paying off big for them right now and they see that as a way of making a lot more money fast." He paused. "Since Leon is a descendant of Hitler just like Hugo and has visions of conquering the world like Hugo did, we're definitely going to be dealing with a repeat of what we dealt with previously."

"All the ingredients are there again," Judy said. "Leon has Hitler's genes, and the new members of Belzex have a tremendous amount of money, the greed to get more, and the capability to carry out their plans."

Todd nodded. "They have access to a vast amount of money, and they can provide all the advanced manufacturing capability and AI expertise they need. They're able to recreate super advanced spying and weapons facilities like Hugo had and, of course, Leon had before."

Matt glanced at Kyle and Ben, "Will you be able to continue to get a lot of support from the other intelligence agencies, and especially the CIA and NSA as we go forward?"

"I see no reason why not," Ben said. "They've been very helpful so far and I expect that to continue."

Ralph looked at his watch. "Let's take another short break to check on some other business matters. We'll keep all we've discussed in mind and we'll come back in thirty minutes and decide our best course of action in going forward from here."

*****

Back in their private jet and heading toward Mexico City, Leon gathered everyone together in a specially designed conference area.

"That didn't work out well for us," Ivan said.

Leon gave a sarcastic laugh and didn't comment.

Jorge leaned forward. "Well, I guess everything is clear. We've got another mountain to climb now. We must kill Baker and Gibson for sure before we can get the land. It looks for certain there's no way they're going to sell, no matter how much we offer for the land."

Bolivar nodded. "Yep, we need to put Tuxtun Industries under new ownership, and we all have plenty of assassins to do that. We just need to have a better plan next time." He paused.

“I’m used to dealing with problems like this in my battles with rival cartels when I’m trying to eliminate their leadership.” He slowly scanned the faces around him. “I have a few suggestions we might want to discuss when we’re ready to put a plan together.”

Leon glanced at his watch. “We have an hour before we land in Mexico City. I think we have time right now, and I’m anxious to get rid of those two bastards.”

Ivan looked at Bolivar. “Do you have some good snipers who can pick them off from a long distance?”.

Bolivar flashed a big smile. “I have a lot of them. They not only are well trained, but they’re also highly experienced. They’ve been very busy for a long time in our wars with rival cartels.” He looked at Leon.

Leon tightened his lips and gave a slow nod. “That will likely be necessary, and we need to discuss every aspect of that in great detail, but it will be best to wait until our next meeting for that discussion, so we’ll have enough time to cover all the possible angles.”

There was a brief silence.

Everyone glanced at Leon waiting for him to make some more comments.

Leon brought up several options for everyone to be thinking about. The remaining discussion took up the rest of the hour.

“After the plane landed, and before they stood to depart,” Leon said. “We’ll head over to Jorge’s house, catch our breath and then determine what our next move is. We’ll think things over and form a plan before we all go back home.”

Chapter 25

After the thirty-minute break, the group in the boardroom at Tuxtun continued their discussion.

Ralph gestured to Ben. “I think the best way to wrap up this discussion is to start with you and Kyle giving some highlights from your latest intelligence on some of the subject matter we’ll be dealing with.”

“Our combined intelligence agencies have accumulated a lot more evidence that the Hidden Empire is still out there and functioning even better than before,” Ben said. “A lot of the information that has been gathered pertains to, quantum computing, AI, and global cashflow in the Specter Matrix.” He paused. “I think Kyle has a lot of new data we need to know about.” He gestured toward Kyle.

“Todd and I have worked together on a lot of the new quantum computing information that has been collected,” Kyle said. “And we’ve both worked with Judy on the global cashflow.” He took a deep breath and exhaled. “We’ll try to explain the significant items in a way that will make sense to everybody.” He looked at Judy. “You’ve explained a lot about the global financial system to us before, but there are some additional things I know you want to mention. I think that’s the best place to start getting into the details. Once we get a better grasp of the broader financial side of this picture, I think what Todd and I explain will make more sense to everyone."

"The International Acceptance Bank of New York was established in 1921 to provide some necessary adjustments to the system the Hidden Empire was in the process of setting up," Judy said. "That worked well enough for a while because the system didn't need to be too sophisticated back in those days." She paused. "However, The Manhattan Company acquired The International Acceptance Bank of New York in 1929, and some changes needed to be made."

Todd nodded. "In 1930 the powerful people who were making a lot of money from the system they had established decided they needed another enhancement, so they created the Bank for International Settlements (BIS) in Basel, Switzerland. This new bank provided a multitude of new services and promoted cooperation among central banks."

"Next the Federal Reserve created the Federal Open Market Committee (FOMC) in 1933," Judy said. "The first branch of the World Bank, the International Bank for Reconstruction and Development (IBRD) was created in 1944 along with the International Monetary Fund (IMF). Both were for the purpose of rebuilding after World War II. The headquarters for both occupies a square block along Pennsylvania Avenue two blocks west of the White House." She paused. "The other four branches of the World Bank were added later."

There was a brief silence and there were several people making notes.

"In 1955, The Manhattan Company merged with The Chase National Bank to form The Chase Manhattan Bank," Judy said. "This led to more changes."

Todd leaned forward. "The International Finance Corporation (IFC) was added to the World Bank in 1956, and the International Development Association (IDA) was added in 1960. The IFC is the private-sector arm of the World Bank and gives it a lot more freedom for money distribution among private entities."

"The Institute of International Finance (IIF), which is a separate entity from the World Bank, was created in 1983," Judy said. "The fifth branch of the World Bank was added in 1988. That was the Multilateral Investment Guarantee Agency (MIGA)."

"The World Trade Organization (WTO) was created in Geneva, Switzerland in 1995," Todd added. "The Chase Manhattan Corporation merged with J P Morgan & CO in 2000 to form J P Morgan Chase & CO." He scanned the faces around him and shrugged.

"That's the overall structure we're looking at," Judy said. "And all those changes contributed heavily to the complexity we're having to deal with now. The billions of transactions daily among the various entities are intertwined in a complex manner." She held a hand, with the palm up, out to each side. "I hope that makes it clear as to why the Hidden Empire doesn't need any shell companies, and to how the Specter Matrix can be so effective."

"I'm sure we've all learned something," Matt said.

Ralph laughed. "That certainly gives us a lot more to think about."

Ben nodded. "It's a lot more to think about, but we'll all benefit from the knowledge."

Ralph's gaze took in both Judy and Todd. "Do you have much more to tell us?"

"A little more, and then we'll wrap it up," Judy said. "Most of the additions to the global monetary system over the years had a lot to do with investments and credit. Beginning with the creation of The International Acceptance Bank in 1921, a lot of money could be made by the owners of various aspects of the global monetary system by charging fees to guarantee loans."

Matt gave a short laugh. "As you mentioned earlier, charging fees to guarantee loans was recognized by the creators of the Federal Reserve System as a quick way to make more money. When they created the International Acceptance Bank of New York in 1921, they started taking advantage of that."

"When the group of super wealthy people on Jekyll Island first envisioned setting up the Federal Reserve they somehow overlooked that way to make money or they couldn't easily incorporate it into what they set up in 1913." Judy gave a slight shrug. "Anyway, they eventually made the adjustment."

Ralph leaned forward. "I think we can be certain the Hidden Empire will have more involvement in our future activities, and we know they will take some drastic action when anyone threatens their operation. It's good to get some more knowledge that will help us stay more aware of how we might affect them."

"Yeah, it's good to remember what Judy and Todd just explained," Matt said. "We should definitely stay more aware of a lot of possibilities for conflict with the Hidden Empire and with Belzex too."

Ralph nodded. "If we continue to grow crops in Mexico we're almost certain to have problems with cartels that use parts of the same system, which means we'll have problems with Belzex." He scanned the faces around the table. "We'll be prepared for a lot of conflict."

"The FBI will support you every step of the way," Ben said. "But what we don't want to happen is to have problems with the Hidden Empire. So, we want to be as careful as possible to not do something that will cause them to want to get us out of their way by eliminating us."

Everyone nodded.

"In all of our research," Judy said. "Todd, Kyle, and I have concluded there's a possibility the Hidden Empire is getting a large money flow from the illegal drug business. If they are, we could step on their toes with what we do contending with Belzex and certain cartels."

Ralph regarded Judy carefully. "Growing poppies is the common ground in this maze of possibilities. That's a major reason for you, Todd and Kyle to continue your research."

"I agree," Ben said. "We need to continue to do our jobs to

the best of our ability, but that doesn't mean we can tolerate being stupid. We need to be as smart as we can be in every situation we find ourselves in."

"Yep, we need to keep a basic understanding of the full environment we're maneuvering around in, financial and otherwise," Ralph said. "If we're careful about not becoming a problem for the Hidden Empire in any way, we should be able to survive and solve the problems we need to solve."

"We don't plan to work toward changing anything in the global monetary system," Ben said. "Any necessary changes will be someone else's job."

Everyone gave a thumbs-up.

There was a brief silence.

"And again, we need to be well organized going forward," Ben said. "It's obvious we're in dangerous waters."

"Do you think we'll hear from Leon Schulz again real soon?" Ralph asked.

"I don't think so, but I think he's more motivated than ever to kill you and Matt," Ben said. He made a wide sweep with his arm. "And probably everyone else in here."

Matt narrowed his eyes. "I'll bet he's planning right now on how he's going to do that. So, we might hear from him again soon."

Most everyone gave a slow nod.

Ben's glance took in both Matt and Ralph. "I think he'll be as relentless as he was before and as Hugo Wagner was."

"I don't think there's any doubt about that," Ralph said. "We know we need to be as prepared as possible."

"Yeah," Matt said. "I vividly remember every detail of all the things that happened when Hugo was trying to eliminate us." He looked at Ralph. "I'm sure you do too."

Ralph nodded. "Yep, every detail. You don't forget any of those types of things."

Ben leaned forward. "I think you're going to encounter a lot of that again. Probably we're all going to encounter a lot of it. We all had better be prepared."

"And that's a reminder we all need to be in top form at all times," Kyle said.

Ralph looked at Ben. "When do you think we need to go back to the practice range?"

Ben laughed. "Funny you should mention that. I've reserved a practice session for you and Matt at nine o'clock tomorrow morning. If you can't make it. I'll schedule another time real soon."

"We can make it," Ralph said. "Staying alive is our number one priority and will continue to be." He looked at Matt.

Matt gave a thumbs-up.

Chapter 26

In New York City, seven elderly men sat in cushioned chairs in a private lounge inside the corporate headquarters of a large global corporation.

Zeus, Jupiter, and Odin, sat with Vulcan, Neptune, Apollo, and Thor.

"A lot of new potential threats are developing," Zeus said. " Jupiter, Odin, Thor, and I thought it would be good to take a bit of a tour and have a joint discussion in each global region." He glanced at Vulcan. "We thought we would start with your group."

Vulcan nodded. "Are you expecting any of these potential threats to fully develop soon?"

"That's what we're going to talk about," Zeus said. "We'll try to reach a conclusion about the things we need to pay the most attention to as we go forward."

Vulcan, Neptune, and Apollo all nodded.

"Remember we still want a one-world government and a religion that controls the population," Zeus said. "That's the long-range goal, but we're not rushing toward it." He paused. "We eventually plan to abolish private property, but we've decided to move forward on that very slowly." Zeus smiled. "I've mentioned before that sometimes these types of discussions get very complicated, so be prepared for that."

Everyone nodded.

Odin studied Vulcan, Neptune, and Apollo for a moment and then said, "We'll try to come up with some action plans in this meeting but remember we have to modify our action plans at times with certain specific actions."

Vulcan narrowed his eyes. "I know you must take drastic actions sometimes for the protection of our organization. The situation with Hugo Wagner was one of them."

Odin gave a slight nod. "Hugo was useful for a while. The disruption he caused made it easier for us to keep all our dealings hidden from public view. Leon was useful for a while too. We didn't have to kill him, but we had to terminate his operation."

Vulcan and Neptune took a moment to light their pipes.

Glancing toward the three leaders of the Hidden Empire, Vulcan asked, "Do you think Leon learned his lesson or do you think he'll give us trouble again?"

"I think he'll give us trouble again, but he's useful until he crosses the line," Zeus said. "We're monitoring his actives closely."

Jupiter nodded. "I think he'll cross the line again and he'll do it soon."

Odin gestured to Thor. "He has all the connections to the KGB that we need whenever we want to eliminate someone or some group."

"We've existed a long time and we had no technology to use when we were first organized," Zeus said. "Now we have the advanced technology we need, and we must stay ahead of everyone else who's using advanced technology. We must ensure no one interferes with how we use the Specter Matrix."

Neptune took a long puff on his pipe and looked directly at Zeus. "I remember that you told us recently our new quantum computing system, if necessary, can instantly alter the cashflow path we use in the Specter Matrix at each decision point."

Zeus nodded. "At each junction in the flow of money through the Specter Matrix, our system can do a billion tests within one second and then decide which path to take going forward."

"We must always ensure the money we send gets to the right person or organization at the right time," Jupiter said.

Odin glanced at Vulcan, Neptune, and Apollo. "Anyone who interferes with that must be eliminated."

"Hugo got a stern warning for what he was planning to do," Zeus said, "and he continued despite the warning. Therefore, he had to be eliminated. He brought it on himself."

Jupiter nodded. "Leon was toying with doing a lot of the same things, but we were able to eliminate his capability to do it."

"Leon's previous plan to get all the gold in the United States would have disrupted our ability to maintain our flow of money in the Specter Matrix," Zeus said. "Which, as we've explained, we need to continue to function properly so we can operate as The Hidden Empire the way we need to." He paused. "And getting back to Hugo, it should be obvious why we didn't want him shooting missiles into the United States."

Everyone nodded.

"We clearly understand why we must constantly pay all the right people off." Vulcan took a long puff on his pipe "And we know we must pay a lot of them big bucks to steer some important world events in the direction we need them to go. And that all leads back to making sure the Specter Matrix functions properly."

Zeus, Odin, and Jupiter all gave a thumbs-up.

Vulcan studied Zeus carefully for a few moments. "In several of our previous meetings you mentioned the Russian Government is indirectly involved in what we do because we control the Russian KGB and when we need them, they sometimes provide hit squads and heavy weapons, like the missiles we used on Hugo, right?"

"That's correct." Zeus gestured toward Jupiter and Odin. "We'll make sure we keep the three of you well informed as things move along with Leon Schulz."

Vulcan, Neptune, and Apollo all nodded slowly.

"And remember our power stretches a long way back. Remember we had to destroy the Titanic back in 1912 to meet some important goals we had to achieve back then," Zeus said. "Back when some of our forefathers, a group of millionaires and congressmen who wanted to establish a new banking system in the United States first met on Jekyll Island in November of 1910." He paused. "The millionaires were all from New York and they paid for a private train to secretly take them from New York to Jekyll Island. After many long meetings they couldn't get unanimous agreement on going forward in establishing the new banking system we desperately needed to accomplish our goals."

Neptune took a puff on his pipe and leaned slightly forward. "I remember the story well. We had to take some extreme measures to get things to go the way we needed them to be done".

"We all remember the story well," Apollo said. "Some of the powerful millionaires and senators who were opposed to the creation of the Federal Reserve System were onboard the Titanic for the maiden voyage. The Titanic sank on April 15, 1912. Some of those onboard who opposed the creation of the Federal Reserve were eliminated." He paused and narrowed his eyes. "Seven senators were on board who were opposed to the new banking system but only one died. It was the same with the millionaires, only three died."

Vulcan nodded. "All the members of Congress who would be voting to establish the Federal Reserve had been properly warned. With the right messages sent to the opposition, the Federal Reserve Act passed Congress on December 23, 1913."

"It's good you remember those details," Zeus said. "And it will be good to keep the general picture in mind as we move forward and cope with new problems that come up."

Neptune took a big puff on his pipe. "I remember you saying many of the descendants of the founders of the Federal Reserve are still around."

"Yes, quite a few are still around, and some are part of our organization," Zeus said.

There was a brief silence.

Zeus scanned the faces around him. "I'm impressed. It looks like you three have retained well the important information we discussed in earlier meetings. I'm not concerned anymore about needing to reinforce the things we've discussed previously."

Jupiter and Odin each gave a thumbs-up.

"We wouldn't be part of this organization if we didn't have exceptional capability," Vulcan said.

Neptune and Apollo nodded.

Zeus grinned. "We've always been careful to make sure we recruit well."

"In previous meetings you had with us you always emphasized the history lessons you constantly gave us all related to maintaining the level of money flow we need so we can always make the necessary payoffs," Apollo said.

"Right, that has always been the main focus of our operation. Without the constant money flow, we can't function like we need to," Zeus said. "Over the years we've had to do a lot of drastic things to maintain that money flow. Remember, we've even had to eliminate some Presidents."

Everyone nodded.

There was a brief silence.

"How long do you think it will be before Leon Schulz becomes a big problem again?" Neptune asked.

"Like we mentioned earlier, we're keeping a close eye on him and Belzex," Zeus said. "The disruption he causes is still useful, but if he gets out of his lane like he did before, and like Hugo did, we'll eliminate him."

Neptune narrowed his eyes. "Knowing what you know, I was hoping you would give me a good guess on the timeframe."

Zeus grunted. “It’s hard to tell. We’ll just have to stay alert and watch everything carefully.”

There was a brief silence and then Zeus stood, signaling the meeting was about to end.

Zeus gestured toward Odin and Jupiter. “I think we all feel this was a productive meeting and we’ll plan to have a lot more as we go forward.”

“Thanks again for reminding us that drastic actions sometimes need be taken in order to maintain our power,” Vulcan said. “We’ll all keep that in mind as we continue to do our jobs.”

“You’re welcome,” Zeus said. “I’m relieved to know you retain the information as well as you do.” He paused. “We must always make sure we will be able to control important events in the world when we think it’s necessary. Jupiter, Odin, and I always try to keep our members fully informed and fully aware of the reason drastic measures need to be taken at times.” He made a fist with his right hand and held it in the air. “To continue our power, we must have the means to remain masters of stealth and titans of influence. We must always do what is necessary to maintain our money flow.”

Vulcan, Neptune, and Apollo all gave a thumbs-up.

Chapter 27

In Mexico City in the plush lounge in Jorge's mansion, Leon scanned the faces around him.

"We must get our hi-tech datacenter ready to use as soon as possible," Leon said. "We can't move our plans forward without it."

"Do you have any suggestions as to how to speed things up?" Ivan asked.

"Yes," Leon said. He looked at Ivan and Jorge. "The two of you have large manufacturing companies." He shifted his gaze to Chen. "You have a large advanced-technology company."

The three men nodded and made no comment. Obviously waiting for Leon to continue.

Leon glanced at Ivan and Jorge. "First of all, you two should increase the workforce you have for constructing the datacenter."

Jorge and Ivan both nodded.

Leon included Chen in his gaze. "The three of you should combine intellectual resources, set up a joint team to have constant discussions, and plan to move forward at a rapid pace.

We want to rebuild a sophisticated structure like we had before and be able to use a lot of advanced AI products easily."

"Have you thought developing any new products that we don't already have in the warehouse?" Ivan asked.

"I've thought about it quite a bit," Leon said. "We need to expand our surveillance capability so we can spy on the Hidden Empire. We want to be more capable than they are and be able to prevent them from eliminating us." Leon narrowed his eyes. "As we get more powerful, I'm sure things will get to the point where they'll try again. And this time I think they'll try to blow us all up like they did Hugo and his operation."

There was a brief silence.

"Do you want us to make any improvements in the hi-tech datacenter from what we had last time?" Chen asked.

Leon nodded. "Remember I discussed Hugo's control center with Jake Bolton many times and I took a lot of notes. I studied the notes a lot, and I made a lot of enhancements. Now, I'm thinking of a few more enhancements. They're a bit complicated. I'll go over the details a little later."

Chen leaned forward. "It might be best for you to tell us now before we setup our team to start the discussions. At least we'll know what your goals are. That will make it easier to design and manufacture the specific products we need."

Ivan nodded. The sooner we know exactly what we want to end up with the easier it will be to design the product and to construct the manufacturing process."

Leon took a deep breath and exhaled slowly as he removed a small notepad from a pocket. He flipped it open and scanned a few pages.

"I like the old-fashioned way best for keeping these types of notes." Leon looked up at the people around him and grinned. "I love hi-tech, but sometimes I like using a note pad." He shrugged.

No one commented.

"Of course I want the latest features in AI technology." Leon looked at Chen. "I'll rely on you to lead that effort."

Chen nodded and didn't comment.

Leon looked at Ivan and Jorge. "And I'll expect you to ensure we're using the most advanced manufacturing processes available."

Ivan and Jorge nodded.

Leon proceeded to give a lot of details.

"A lot of what you're talking about overlap in some areas," Chen said, "but we'll work all that out when the team we assemble has their detailed discussions."

Leon gave a thumbs-up and then glanced at Bolivar. "We have a tremendous amount of money right now and a lot of it is from your operations. We'll have no delay getting started, but we'll need a lot more soon to pay for all we must do to overcome the Hidden Empire and dominate the planet."

"It's going to be a problem not getting the extra income from increasing our production of Poppies," Bolivar said. "Not getting the extra land will be an ongoing issue. We can increase our production a little, but not much if we don't get the extra land." He paused and stared at Leon for a few moments. "So, I guess it's clear what we must do."

"We'll get the extra land." Leon clinched a fist. "We have a lot of topnotch assassins, and we'll make sure Ralph Gibson and Matt Baker don't live much longer."

Everyone raised a fist in the air.

"We'll insure we have the best AI on the planet along with the best overall technology, including weapons systems," Leon said. "We'll need a tremendous amount of funding to get more capability in AI than the Hidden Empire has."

Chen nodded. "To even use the AI capability that we develop we'll need a lot more energy than we have access to now. We'll need to build a lot more powerplants." He looked at Ivan and Jorge. "That will, of course, be projects for both of you to concentrate on."

"We'll plan on getting new power plants built and operational as soon as possible," Ivan said. "We collectively have a lot of available land around the globe, and we'll select the best sites and get started soon on the work."

Leon glanced at Chen. "I'm hoping your experts can develop some new technology that will improve the building process and manufacturing process quality wise and speed wise."

"We have some new tools we've recently developed to monitor various manufacturing processes," Chen said. "They're valuable for eliminating redundancy and improving every aspect of quality control."

Ivan nodded. "We've developed new tools along that line also. When the members of the new team we're going to form get together we'll use some new methods to select the most appropriate tools that we have available."

"Maybe you can improve the tools you select even further," Leon said. "Don't put any boundaries on what you can do. Remember we want to have more capability in everything than the Hidden Empire does. And we want to get to that point as quickly as we can."

"That's certainly our goal." Chen glanced at Leon and then looked at Ivan and Jorge.

Everyone nodded.

"We already have a lot of tools to monitor all the manufacturing processes and eliminate redundancy," Ivan said. He scanned the faces around him. "Most people would be surprised at how much redundancy there is in the older manufacturing processes. Jorge and I would welcome any additional improvements that become available."

Jorge nodded.

"Okay, it sounds like we're all thinking along the right lines," Leon said. "Let's not waste any time and organize our next steps immediately."

After another hour of discussion, Leon said, "I think we're well organized, let's get ready to get busy."

Bolivar glanced at his partners. "I'll do everything I can to increase our revenue from my current drug trafficking operation. We're doing well processing the poppies from the land we have. I hope we can get more land soon."

"And once we get the extra land, I'll make sure the extra income we receive from the drug trafficking will be put to good use," Leon said.

Chen nodded. "We want to be able to immediately fund every great idea my engineers come up with."

"We'll gain on the Hidden Empire quickly," Leon said. "If our plans come together properly, we'll be powerful enough very soon to protect ourselves from them in every way that's necessary."

Chapter 28

The next day, at the FBI Headquarters in Washington DC, Paul Becker, the FBI Director, sat in a large conference room with several other people, including his counterpart from the CIA, Joe Frank. The room was a makeshift operations center designed to facilitate the two intelligence agencies working together.

The group from the CIA included Craig Parker, head of the terrorist threat investigation unit, Wendell Jones, technology specialist who reports to Craig Parker, and two field agents, Kevin Brown and Butch Reilly.

Included in the group from the FBI were Gary Stennis, head of the special investigation unit for domestic threats, and Dan Maxwell, a technology specialist who reports to Gary Stennis.

Paul Becker scanned several rows of monitors along two walls. He shifted his position slightly and drummed his fingers on the conference table as he then focused a steady gaze on Dan Maxwell, his technology specialist. “So, you’re keeping Belzex under close surveillance?”

“Yes sir,” Dan replied. “Wendell and I constantly stay

closely coordinated on that. We have the latest advanced technology and we're watching everything they do."

Paul nodded as he scanned the faces around the table. "I got a call a couple hours ago from the NSA Director. They have a lot of new advanced capabilities, and he gave me some new information they had just uncovered." He paused. "I thought, since we're all involved in an investigation with a lot of loose ends, we should compare notes among ourselves before we go forward with all the other branches of law enforcement."

Everyone nodded.

Paul looked at Dan and Wendell. "Have you found any new information we should all know about?"

"Belzex has already set up a lot of new dummy corporations to hide their current financial dealings, all illegal for sure," Dan said."

"Corporations set up in the Channel Islands?" Joe asked.

"There and on the Isle of Man and a lot of new places," Dan said. "The British Virgin Islands, Bermuda, Switzerland, Singapore, and Hong Kong are now in the group." He went on to give more details.

Paul leaned forward. "Since Belzex is again establishing some type of joint office in the United States, and it looks like it will be back in Jacksonville, Florida this time."

"Wendell and I are in the process of investigating it," Dan said.

Joe nodded. "At the CIA we think the purpose of the office will again be to facilitate and closely coordinate the activities of the billionaires at Belzex with these dummy corporations and with key players in the global monetary market. There are a lot of details. We'll discuss more about that later. Right now, I suggest we concentrate on what we think these billionaires in Belzex will be doing next with the huge amount of funding they plan to receive."

Paul glanced at Joe. "I would like to pause a moment and see if anyone has any questions they think need to be answered about the joint office before we continue with the broader discussion."

"I have a question for Dan that I think needs to be answered before we get too far along in the discussion." Craig looked at Paul.

Paul motioned for him to go ahead.

Craig looked at Dan. "Who do you think are the key players in the global monetary market that might be funding this office?"

"I think the IMF, the WTO, and the World Bank are the key players globally, but Wenell and I haven't found out yet if any of them are involved" Dan said. "Remember the World Bank has five different divisions, and we're investigating all of them." He paused. "Wendell and I have found very few details so far about the source of the funding for this joint office, but we've found a lot of clues."

Wendell nodded. "So far, it looks like none of the funding comes directly from any part of the World Bank, the IMF, or the WTO, but it could be coming indirectly from them." He went on to explain.

"Even though we've found no evidence of where the funding comes from, we can make some good guesses," Creig said.

"And….?" Paul asked.

"We think the funding for the joint office is sent through a complex mix of accounts in some of Belzex's dummy corporations," Wendell said. "This mix of accounts in the dummy corporations is connected to a variety of accounts in the global monetary institutions. We'll explain more later, it's complicated."

Dan gave a quick nod. "It's complicated, but our advanced technology has allowed Wendell and me to unravel a few things, and we'll continue our research and unravel a lot more."

"Why do you think they again felt a need to establish this joint office?" Joe asked.

"We're sure it's the same purpose as before," Wendell said. "And we think various aspects of all the global monetary organizations are again being used by Belzex to secretly funnel money to activities and organizations they want to support. Having an office with specialists from each of the major global monetary organizations and the shell companies makes things easier in a lot of ways."

"We've also found evidence of both the Kremlin and Beijing being heavily involved in some of the recent activities of Belzex," Dan said. "Wendell and I expect them to stay heavily involved."

Wendell leaned forward. "We're guessing certain elements of the WTO, the IMF, and the World Bank in this joint office will also be working closely with our Federal Reserve and they probably plan to have a lot of face-to-face meetings to avoid their phone calls being intercepted. Like before, having an office over here makes things a lot easier for them."

Paul nodded. "NSA has improved their technology. Phone calls can now be intercepted and decrypted no matter how good the encryption is." He glanced at Dan and Wendell. "You should be getting a lot of useful information from them."

Dan and Wendell both nodded.

"NSA is a big help and we're staying in close touch with them," Dan said.

"They have over 32,000 employees, and most of them are working to develop emerging cryptographic tools, and advanced cybersecurity AI systems." Wendell scanned the faces round him. "They constantly employ new mathematical breakthroughs in an attempt to ensure a decisive technological advantage." He paused. "Dan and I both think they're successful at what they do."

"So, it sounds like they're constantly a big help to you, right?" Joe asked.

Dan and Wendell both nodded.

"I didn't know Switzerland had a lot of shell companies. Can we take a few moments for you to tell us some more about that?" Kevin asked.

"There are over 33,000 in Switzerland and Dublin has

almost 14,000 of them," Dan said. "They'll not all for illegal purposes, but many of them are."

Wendell nodded. "We've discovered three that Belzex is using, and they might be ones that are being used to fund the new joint office being established in Jacksonville."

After a slight pause, Dan said. "Two of the billionaires in Belzex own manufacturing companies and one billionaire owns a company that specializes in AI products."

"Looks like Belzex has the same capabilities as before," Paul said. "I think we can expect them to get back on the same path they were on last time."

The discussion continued for another hour.

"We have a lot of new information to absorb," Joe said. "After we finish this discussion we need to make sure the group in Jacksonville is fully informed. It looks like we'll need to give them a lot of support."

Paul nodded. "I'll give Ben a call later today." He gestured toward Dan and Wendell. "I suggest the two of you plan to go down there in the next few days. I'll make sure Ben is expecting you. It will be good for all of us in the Washington DC area to coordinate more closely with those in Jacksonville as we go forward."

Everyone gave a thumbs-up.

Joe glanced at Paul. "I have a feeling some big things are going to be happening soon that will affect both of our organizations. It's good for each of us to have a person down there who can interact firsthand with the necessary people as things are developing and ensure that we get the correct information up here in the Washington area."

Everybody nodded.

"We know this situation with Belzex is potentially a big threat to us and it's probably going to get worse quickly," Joe said. "We can't just sit around and wait for something to happen."

"We must stay on full alert at all times and guard against a lot of possibilities." Paul rubbed his chin. "Having the Hidden

Empire lurking in the shadows doesn't make things any easier for us. But there could be a positive side to that. Belzex will likely continue to improve their capability and interfere with the operation of the Hidden Empire's Specter Matrix."

"I'm expecting the Hidden Empire to soon eliminate Belzex like they did last time," Joe said.

Everyone nodded.

"As the NSA continues to improve their technological capability, and especially the technology they use for surveillance, the Hidden Empire will probability come after us too, " Joe said. He scanned the faces around him. "We had better be prepared for that."

Everyone continued to nod.

Chapter 29

The next morning in Jacksonville, Ben Fulton and Kyle Rankin sat at the large table in the main conference room in the FBI headquarters on Gate Avenue.

Nine other people sat at the table with them.

Ralph Gibson, Matt Baker, Judy Caldwell, Todd Green, and Steve Baxter were there from Tuxtun. Justin Mason, the Jacksonville Police Chief, and Vince Simmons, the Saint Augustine Police Chief were there.

Dan Maxwell from the FBI and Wendell Jones from the CIA were also there from the Washington DC area.

"We have some new information to discuss and much of it is about the Hidden Empire and their use of the Specter Matrix, which means a lot of the discussion will involve the use of quantum computers." Ben gestured toward Dan and Wendell. "Two of our top technology experts can get us started."

Ben looked at Todd. "And I'm glad you could attend. Going forward, you and Kyle both need to work closely with these two guys." He again gestured toward Dan and Wendell.

Kyle and Todd each gave a thumbs-up as Dan and Wendell

were getting started with the discussion.

After Dan and Wendell had taken turns providing key information about how quantum computers were being used by the Hidden Empire and by Belzex, Ben said, "We don't know for sure, but we think the Hidden Empire and Belzex are doing similar things with their computers and their money flow." He paused. "Which means they're both using quantum computers to control their global cashflow."

Everyone nodded and made no comments. Matt figured that was because they wanted to hear more about the subject matter before they started asking questions.

"We know Belzex uses a lot of shell companies as part of their cashflow system, and the Hidden Empire doesn't use them because they don't need them." Ben looked at Matt and Judy. "We'll need both of you to explain more about that a little later."

Matt and Judy both nodded and didn't comment.

Ben leaned forward. "We have a lot of challenges coming up and we're going to be working closely with the Defense Advanced Research Projects Agency (DARPA), the Intelligence Advanced Research Projects Activity (IARPA), and the Homeland Security Advanced Research Projects Agency (HSARPA). We'll discuss all the details about this later. We're all going to work on multiple matters at once along several intelligence channels." He glanced at Ralph and Matt. But, as before, most of your intelligence work will be with the FBI, CIA, and DEA. I just don't want you to suddenly be contacted by an agency you aren't familiar with."

Ralph and Matt each gave a thumbs-up.

Dan looked at Todd. "If you have time, Wendell and I plan to get you more involved with RAND. They've accumulated a massive amount of data on a multitude of specific subjects we're analyzing. "

"The experts at RAND think there's going to be a future problem with the world's supply of gold, and they think Belzex is connected to that," Wendell said. He explained.

Ralph rubbed his chin for a moment and then leaned toward Ben. "How does this tie into what we're going to be dealing with in the next few days?"

Ben gestured to Dan.

"NSA has intercepted some more calls among these foreign billionaires wanting to buy land in Florida," Dan said. "They mentioned planning a gold heist several times in their conversations."

"Involving Fort Knox?" Matt asked.

"They mentioned Fort Knox, but they talked more about the gold in the Federal Reserve vault in New York," Dan said. "It has more than Fort Knox." He paused. "The US Mint has 7,500 metric tons of gold but it's not all at Fort Knox, only 4,500 metric tons is in the vault at Fort Knox. The Federal Reserve Vault in New York City has 5,000 metric tons."

Matt looked at Dan. "Since the Federal Reserve keeps their gold in a vault that's eighty feet underground and on a big slab of solid granite, I assume Belzex is going to try something similar to what they've tried before?"

"It's different this time, the leaders at Belzex have a slightly different way to get it this time. They're going to let the Federal Reserve Bank do everything for them," Dan said. "From the conversations NSA intercepted and decrypted, they're planning to receive a large number of shipments from the vault with dummy transactions they can disguise as being legitimate. They control numerous dummy corporations they can use to pull that off. The foreign billionaires plan to take advantage of the constant number of transactions for gold bars and gold coins to be taken out of storage. My understanding is that some are removed from the vault on a daily basis."

The discussions continued for about a half hour on the details of the scheme Belzex was planning to execute.

"We'll see how things play out," Dan said. "Wendell and I will keep everyone up to date." He looked at Matt and Judy. "If you're willing, I hope you'll work with us on some of the financial specifics."

Matt and Judy nodded.

There was a brief silence.

"Our work is taking on a completely different look in a hurry." Ben's gaze took in both Ralph and Matt. "If you continue to work closely with us on all this, the Hidden Empire might think you're sticking your nose where it doesn't belong. For your self-protection, I suggest you sharpen your combative skills immediately."

Vince scanned the faces around the table. "That actually applies to all of us. We're all going to be in increasing danger as things develop."

Ben looked at Vince. "Is there any specific danger right now you're thinking about?"

"I think all of us are in danger of being shot in the near future," Vince said. "I'm sure these foreign billionaires in Belzex have lot of assassins lined up to take care of anyone who creates problems for them. For one thing, I'm guessing they have a lot of expert snipers available."

"Any guess about how soon someone might attack us?" Ralph asked.

"I think Belzex will target you and Matt again real soon. I don't know about the Hidden Empire," Vince said. "I'm guessing Belzex is desperate to buy the land to grow poppies on, and they need to get you out of their way. They probably have some very expensive new projects they're starting, and they need the additional money flow from the land."

Ben leaned forward. "Remember, the Hidden Empire has access to the Russian KGB, and they have basically an unlimited number of assassins. They might decide to take some action too against us too. I'm sure, with all the advanced technology NSA is sharing with us, we could all be viewed as stepping on their toes."

There was a brief silence.

Judy scanned the faces around the table. "I suspect Belzex is starting some very expensive new projects with AI, and they need a huge amount of money to do what they need to do." She glanced at Ralph and Matt. “Belzex needs a huge amount of money to build more powerplants to produce a lot more power for their expanded AI usage. I think you can be positive Leon will try to eliminate you, and soon.”

Matt narrowed his eyes. “Yeah, he’s probably setting up a high-tech control center again and providing the necessary energy for that is going to cost some real big bucks.”

“And he’ll probably try to conquer the world like Hugo did,” Ben said, looking slowly around the room.

Matt tightened his lips. “Yeah, I think he’s going to put a lot of effort in preparing to do that. Being a descendant of Hitler, he can’t resist. It’s in his genes.”

Ben looked at Ralph and Matt. “I think we can be certain he’s highly motivated to continue to pursue his goal of conquering the world.” He continued to look at Matt and Ralph. “You’d better start polishing all your defensive skills right away.”

Ralph grinned. “I don’t think you need to try too hard to convince us to do that.”

Matt nodded.

Ben leaned forward. “Good. We’ll get both of you scheduled right away.”

“I’ll authorize you to use our police practice range in Saint Augustine,” Vince said.

Justin gave a thumb-up. “I’ll do the same here in Jacksonville.”

“I’ll also make our FBI practice facility available for all of us,” Ben said. “We all need to be as well-trained to protect ourselves as much as is possible.”

Ralph scanned the faces around the table. “There’s no need for anyone to try to persuade us. Matt and I will work on polishing all aspects of our self-defense capability right away. We’ll expect a variety of danger, and we’ll constantly improve our ability to cope with any of it.”

“We want to keep on living, and we definitely want to keep all of our defensive skills finely tuned,” Matt said. “We’ve had many confrontations over the last few months and, speaking for myself, being able to draw my Glock and fire within a half second has proved to be extremely valuable for quite some time now."

Ralph nodded. "Ditto."

Everyone gave a thumbs-up.

Ben leaned toward Matt and Ralph. "I also recommend we schedule some more hand-to-hand combat training with the toughest instructors we have. The way things are shaping up that might be something that becomes more valuable as time goes on."

Matt and Ralph both nodded.

Chapter 30

Leon Schulz sat in his study in his elaborate home in Germany and prepared to ask Jake Bolton some more questions about Hugo Wagner's technology setup.

"This is all super important," Leon said. "So, I want to again go over some of the things we've talked about before."

Jake nodded and then answered a lot more questions and again gave a detailed description of the control room Hugo had designed for his purposes.

"The control room was round because Hugo wanted it that way," Jake said. "Oleg and Wan especially seemed to be okay with it. I never heard any arguments from anyone."

Leon nodded slowly and then looked at his notes. He grabbed a pen and looked at Jake. "Give me the exact details again."

"The control room was about fifty feet in diameter, with curved computer screens which extended from floor to ceiling. The ceiling in the control room was about twenty feet high, and the multiple displays were intermixed from floor to ceiling," Jake said. "The table had a circular desktop with two feet of desk space extending all the way around the doughnut hole in the center which was twenty feet in diameter."

Leon nodded constantly as he added additional information to his iPad.

"A small section of the desktop could be lifted to allow people into the center area," Jake said. "Four hi-tech chairs sat in the open area in the center portion. The chairs could be moved around easily to different parts of the desktop, and they had a 360-degree rotation."

"All the displays were controlled by a quantum computer which was connected to thousands of surveillance devices (both video and audio), right? Leon asked.

"Right, and don't forget all the people watching the displays could control selected ones in various ways by using their hi-tech remote device."

Leon studied his iPad for a few minutes and made some more notes.

After some more discussion, Leon said. "Okay, I think I'm set for now." He stood and looked at Jake. "Thanks."

Jake gave a thumbs-up.

"I've carefully reviewed everything to make sure I didn't miss anything important when we discussed this before," Leon said. "When our new team finishes their preliminary planning on how they're going to get started with all the manufacturing processes, I'll have a coordination meeting, and we'll discuss more details on how we'll build the new hi-tech datacenter. We'll discuss both the design of the center and the design of each of the new products."

Jake nodded. "Are you having another meeting with all the Belzex group right away?"

Leon looked at Jake and didn't immediately comment. He was mulling a lot of things over carefully.

"Well, what are you going to do?" Jake asked. "You looked like you were thinking hard about some things."

"There are a lot of things to consider." Leon narrowed his eyes and looked down at the floor for about a minute and then he slowly lifted his head and looked back at Jake.

"So, when's the next meeting?" Jake asked.

"Very soon," Leon hesitated for a moment and then started slowly rubbing his chain.

Jake didn't comment. He patiently waited for Leon to say something.

"You'll need to attend the meeting," Leon said. "There will probably be some questions that only you can answer. Some questions will probably relate to how Hugo decided to use certain parts of his AI capability and the reasoning behind those decisions."

"I might not have all the answers, but I'll be glad to help in any way that I can," Jake said.

There was a brief silence.

"We should have all the AI capability we need to move forward with our plans if we build our hi-tech datacenter to specifications and have it fully operational." Leon paused and looked at Jake. "I think this time; one will be enough. We know we can't hide from the Hidden Empire, so having a backup isn't necessary. They'll know where each one is and when they think they need to, they'll come after us wherever we are."

Jake narrowed his eyes and turned fully toward Leon. "We'll put a lot of effort into defending it.

"Yeah, to compete with the Hidden Empire, we must have a lot of extra capability, and we'll need it as soon as possible," Leon said. "We know we won't get too far along in our plans to conquer the world before the Hidden Empire comes after us. We must make sure we're ready."

Chapter 31

Just off the coast of Georgia at a remote location on St. Simmons Island, Zeus, Jupiter, and Odin stood together in a corner of the ornate lobby in the building where the next meeting of a prestigious global society was about to take place.

They had received invitations to the meeting under their real names and were highly regarded among the global elite.

Zeus knew they all thought it was good to attend this meeting every year to make sure they were keeping up with important issues that could affect them.

This meeting also gave them an opportunity to influence a lot of powerful people in ways that were beneficial to the Hidden Empire.

As they prepared to enter the auditorium, they had a final discussion about their strategy.

"We want to make sure we're keeping up with who the most influential people are, and we'll mill around at the appropriate time and discuss various subjects with them," Zeus said.

Odin and Jupiter each gave a thumbs-up.

"I'm sure we're planning on applying the usual strategy," Jupiter said. He leaned toward Zeus. "While we're discussing things with everyone, we'll concentrate on being matter of fact. We don't want to come across as being too anxious about something."

"Right," Zeus said. "We'll be careful about the impression we make."

Odin glanced at his two companions. "We're well experienced in this. We do want to ask a lot of questions to the most powerful people, but we need to be on our toes and be aware of when to stop."

"Right, we have to apply just the right balance of interest, concern, and curiosity," Zeus said.

Odin and Jupiter nodded.

"In past meetings I've noted some of the most powerful people don't want to talk a lot and they certainly don't want to answer a lot of questions," Jupiter said.

Odin looked at Zeus. "That's been my observation too, but we have the skills to get the information without pressuring them,"

"We can be extra careful with some people and still find out what we need to know." Zeus glanced back at Odin. "You're correct, we have the necessary skills."

A brief silence ensued.

"Some people are very guarded on some subjects. Some seem to get upset when you even mention certain things," Jupiter said. "During some conversations I've had, I've quickly realized I needed to back off."

Zeus gave a short laugh. "Yep, sometimes you need to do that." He paused. "But in some cases, it might be good to use a different technique and keep trying. When dealing with some subjects we need to make sure nothing has changed in some critical areas that will affect us."

Odin and Jupiter gave a thumbs-up.

"Remember, we have the skills to collect all the information we need," Zeus said. "Sometimes we just need to try a little harder and occasionally a lot harder. We certainly need to make sure we're aware of how well-known the new billionaires in Belzex are, and we also want to know how well-known Belzex is now that's it's been rebuilt."

Jupiter nodded. "That will help us monitor if Belzex might become a threat to us again or not."

"And it will help us determine which of the influential people in this meeting we'll need to stay in close contact with," Odin said.

Zeus nodded. "We must try as hard as we can to get an accurate picture of where we need to spread more money around to make things go the way we need them to go. Many of the people attending this meeting will have some involvement in that."

"Yeah, we must keep up with where the big pockets of money are globally and who has the most needs for more money," Odin said. "Attending these meetings always gives us a golden opportunity to gain much more knowledge to make the correct decisions."

Jupiter scanned the faces around him. "Odin has a good point. We must be able to make the right decisions on where to spread the money around."

"Yes, to people and to organizations." Zeus rubbed his chin and turned toward the long row of doors that led into the auditorium.

Chapter 32

Matt approached the entrance of a firing range early the next morning that was south of Jacksonville.

It was the one Vince Simmons provided for their use where the Saint Augustine Police Department got in most of their practice sessions.

Matt knew Ralph had the first practice section and he was scheduled next. He was right on time, and he saw Ralph coming toward him.

"How did it go?" Matt asked.

"I did well," Ralph said. "I had a little rust to shake off, but I did that quickly. I figure you'll shake off the rust quickly too."

Matt gave a thumbs-up. "Glad you did well." He waved to Ralph and kept going.

"See you this afternoon," Ralph called back over his shoulder.

"Roger." Matt knew they had a joint practice session scheduled for later.

In less than a minute, Matt faced a large bluff, over 50 feet high and about 150 yards away. He scanned the area and saw a lot of piles of wood that blended in well with the dirt, same as last time.

"So, you're here for a tune-up like your boss just finished?" The instructor at the pistol range turned fully toward Matt.

"Yes," Matt said. "I'm ready for action."

"Are you still carrying your Glock 19?" the instructor asked.

Matt nodded.

"We have plenty of 9mm ammo for you to use," The instructor said. "I want to go back to the basic training-level and work up. You have any objection to that?"

"No objection at all," Matt gave a slight shrug.

For the next twenty minutes, Matt fired at stationary targets. He hit all of them.

"Very good for not having practiced in a while," the instructor said. He studied Matt for a moment. "Now, you'll execute a fast draw from a holster under your coat and positioned toward the back of your right hip. I think that's the best option. You must be proficient in getting your pistol into action quickly in all circumstances."

"I remember all that," Matt said. "Let's keep going."

The instructor waved his hand in a sweeping motion across the dirt field. "Bad guys could pop up anywhere." He held up his stopwatch. "You need to shoot them as fast as you can, computer will time you. I'll check some of the times."

"Are you still using the cardboard targets for this part?"

"Yeah, they're all cardboard targets," the instructor said. "But remember, they won't all be bad guys. You'll have to judge." He jutted out his jaw. "Don't shoot the wrong ones."

Matt glanced at the instructor. "I won't. I remember this part well."

"Remember, pop-ups will be random, exposure times too." He studied Matt for a moment. "You still look like you're in good physical shape."

"I still exercise regularly."

"Good," the instructor said, "at least you have some good habits. Let's step up the difficulty and see how much rust you've accumulated."

Matt didn't comment.

"Slap in a full magazine," the instructor said, "time to get started on the challenging stuff."

Matt inserted the magazine of fifteen 9mm rounds into his Glock 19. He remembered there was no safety, except on the trigger.

"Before you chamber a round, do some practice draws until you're comfortable," the instructor said. "Don't want you shooting yourself in the foot." He grunted. "I've had that happen too many times before."

Matt had no argument with the logic. He inserted the Glock into the holster and did a practice draw several times.

The instructor ascended a small stairway to a raised platform. Standing on the platform, he looked down at Matt. "When you hear the buzzer, be ready to engage the targets. Any questions?"

Matt shook his head.

"Okay, time to chamber a round."

Matt pulled back the slide on his Glock and heard a round fall into place.

The buzzer sounded.

Matt scanned the area. There, to his far right, a figure. He drew, aimed, but held his fire. It was a good guy. He holstered his Glock and resumed scanning.

Another target, he drew and fired. Got him, it was a bad guy all right. He was sure, even though it was only visible for a couple of seconds.

The instructor stayed on the platform and remained silent.

Continuing the exercise, Matt hit target after target, but he wasn't drawing as fast as he could. He was being a little conservative until he felt fully warmed up. While he was firing live rounds in rapid succession he remembered that little bit about not shooting yourself in the foot.

He carried four extra full magazines, giving him seventy-five bullets. After every seventy-five shots, he knew there would be a break in the action, and he would receive five new magazines.

After an hour of concentrated effort had elapsed. He had consistently hit the targets. He felt a new surge of energy. He knew he was close to being back in top form.

In a couple more minutes, the instructor called a halt and descended from the platform.

Matt holstered his pistol and watched the instructor approach.

"We still have a little work to do." He stared at Matt and mumbled, "At least you didn't shoot yourself in the foot."

"I hit all the targets," Matt said.

"You did." The instructor narrowed his eyes. "But you need to get faster. Based on what I've been told, I would suggest you get back to top speed very soon."

"What were you told?"

"That you're going to be confronted by a lot of world-class assassins, and we need to help you get any accumulated rust off and be as prepared to defend yourself as possible." The instructor stared at Matt for a moment and then frowned. "Based on the types of challenges you're probably going to face, I would suggest you shake off every bit of the rust in one hell of a hurry. We'll work hard to do that."

Matt nodded.

The instructor gave a harsh laugh. "As of right now, I think you have quite a bit of work left."

"Improving my speed on the fast draw?"

"You'll probably find your opposition will be tougher than you've experienced before." The instructor grunted. "Ben Fulton sent a message and told me to make sure you left here in top form."

"I'm warmed up. I'll go at top speed now. I don't think I'll be taking a chance on shooting myself in the foot."

The instructor didn't comment as he turned and rushed back to the top of the platform.

"Okay, get ready." The instructor raised his hand and

brought it down sharply. The buzzer sounded.

Matt alertly scanned the area and responded, hitting all the targets. After he expended his ammunition, the instructor rushed down from the platform

"I was a lot faster this time and I'm sure I got all bad guys," Matt said.

"At least you got the last part right, but you're still too slow."

Matt didn't comment. He knew he was improving fast, no matter what the instructor told him. He figured the instructor was sticking to a script and going through a motivational technique. Matt smiled. His high school and college football coaches used to do the same thing.

"So, get off your ass and get to work." The instructor studied Matt for a moment and then turned and stormed back to the top of the platform.

After returning his Glock to the holster toward the back of his right hip, Matt adjusted his sports coat and again faced the target area. He took a deep breath and then nodded he was ready.

The buzzer went off.

Matt scanned the area. There, to his left, a bad guy. His right hand brushed back his jacket and moved the pistol from the holster with a good sliding motion. He executed a quick forward snap, aimed at the target, and pulled the trigger. The target went down.

He repeated the action with constantly better results. He now knew for sure he had shaken off all the rust.

"You're gotten your speed back to the half second mark. You're done," the instructor said. "I have more students to deal with."

Chapter 33

That afternoon, Ralph and Matt stood inside a large dome, which was the new FBI training facility in Jacksonville.

Each wore a jacket with casual clothes and carried a Glock 19 in a holster attached to the belt toward the back of the right hip.

A replica of a city street stretched before them. Everything was dark except for a few streetlights, good simulation of night conditions.

The instructor swung his right hand in a sweeping motion. "This facility simulates all conditions, including wind and rain. We'll also get some city-street action for both day and night conditions. He studied his two students for a few moments and then asked, "You ready?"

Matt and Ralph both nodded.

"I hope you do better than the other participants. It's a tough course and you need to draw fast. The targets will only be exposed for one second." The instructor narrowed his eyes. "Do you have any questions?"

Matt and Ralph remained silent.

"Let's see how well you can handle this." The instructor

turned and climbed the steps to a platform 30 feet high.

Matt stood on the right side of the street, Ralph on the left side.

The high-intensity streetlights dimmed. Faint silhouettes of buildings loomed around them.

A fine mist fell from the simulated night sky as Matt and Ralph moved forward along the dark street.

Matt scanned the training area, the size of a city block. A target popped up to his right. He drew fast, fired, and hit it.

For several minutes, Matt and Ralph picked off targets while they both moved down the street.

Staying alert, Matt saw targets on both sides of the street. He shot all of them. He noted Ralph was doing a lot of shooting too.

"Hold it," a voice from the darkness shouted. As bright light flooded the area, the instructor hurried down from the platform and over to where Matt stood.

The instructor gestured for Ralph to come running over and then said, "You need to do better or you're going to die."

Both Matt and Ralph stared at the instructor.

The instructor shook his head. "This was a piss-poor effort if I ever saw one."

"What are you talking about? Ralph asked. "I know I got all of them, and fast too."

Matt nodded. "Same here."

"Neither of you saw that last target." Disgust covered the instructor's face. "In real life, you'd be dead."

Matt was suspicious the instructor was just taunting them. He glanced at Ralph. He also looked skeptical.

The instructor continued to glare. "When you start again, move your head from side to side. You can best detect movement out of the corners of your eyes."

"That last target must have been well hidden," Ralph said.

"No more than a typical enemy you would encounter in a downtown setting. Start scanning properly if you want to stay alive." The instructor shrugged. "In this type of situation, keep rotating your head. That might save your life, unless you screw something else up."

Matt nodded. So did Ralph.

"I've heard both of you think you're good, but you'd be dead in a real situation like this." The instructor turned and stomped off, glaring back over his shoulder. "Now let's see some improvement."

After climbing to the platform, the instructor said, "This has been a dismal performance so far."

The instructor was right about one thing for sure, Matt thought: They needed to improve their skill in this scenario.

The lights dimmed.

While turning his head from side to side, Matt moved forward through the fine mist. Determined to improve, he scanned the area. There, to his right, on the rooftop. He drew and fired. The target went down.

He continued for several minutes, eliminating target after target.

Reloading quickly was part of the exercise. Matt carried four extra full magazines. He knew after he and Ralph each fired seventy-five shots, there would be a break in the action. They would each receive five new magazines.

After one of the breaks, the simulation shifted to daytime conditions.

The additional practice lasted two hours before the instructor called a halt and descended from the platform.

Matt and Ralph moved to the middle of the street and watched the instructor approach.

"If you both can maintain the skill level you've achieved, both of you have a reasonable chance to stay alive," the instructor said.

Matt shared the instructor's opinion. He noticed Ralph appeared equally confident.

The instructor studied them for a moment. "It's good you did okay on this part. I'm told you'll need to apply these skills soon. Now go change clothes for the next part."

A few minutes later, Matt and Ralph, each wearing SWAT team gear, stood ready to continue the training.

The instructor handed both a rifle with a short stock and two pistol grips. "This is a SIG 556. It uses NATO 5.56mm rounds and can fire 900 rounds per minute. It has a thirty-round magazine."

Matt held the compact rifle level and juggled it. Ralph did the same.

"It weighs 7.8 pounds." The instructor pursed his lips. "It's exactly 37 inches long. Good weapon to use at night. Flash suppressors disperse the gas venting from the barrel, diffusing its brightness, helps conceal your position. A good marksman can hit a four-inch square target at 400 yards with it."

Matt made no comment. He figured the instructor had a set speech he used every time.

Ralph made no comment either as he nodded at the instructor.

"We're now using this weapon in many intelligence and law-enforcement agencies," the instructor said. "We'll get some more city street action in nighttime conditions and then we'll go to the outdoor scenarios, any questions?"

Matt and Ralph remained silent.

The instructor checked his watch. "Let's see how well you can use it. Set it for semi-automatic fire." After waving them to their starting positions, he turned and climbed to the platform.

Ralph stood on the right side of the street this time, Matt on the left side.

The high-intensity streetlights dimmed. Faint silhouettes of buildings loomed around them. A fine mist fell from the simulated night sky as he and Ralph moved forward along the dark street.

Matt held his silenced SIG 556 rifle in ready position and scanned the training area. A target popped up to his left. He fired and hit it.

For several minutes, he and Ralph picked off targets while they both moved down the street.

Staying alert, Matt saw a lot of targets pop up. He shot at all of them, but he wasn't sure whether he hit them or not. They might have gone down automatically.

"Hold it," a voice from the darkness shouted. Bright light flooded the area. The instructor hurried down from the platform and over to where Matt stood.

The instructor gestured for Ralph to join them.

"Keep performing like this and you'll get your ass shot off." He scowled. "You have a good weapon, use it. That applies to both of you." He turned and stomped off toward the platform while glaring back over his shoulder. "Now let's see some real improvement."

Matt figured the instructor was just following a script to keep them motivated to perform as well as they could. That was okay with him. He figured it was okay with Ralph too. They would need to be as sharp as possible anytime they encountered professional assassins.

The lights dimmed.

While turning his head from side to side, Matt moved forward through the fine mist. Determined to improve, he kept his SIG 556 ready.

To his right, on the rooftop: another target. He fired, got it. Moving forward, he hit target after target. He knew he had hit them because they had all gone down instantly. He had it now.

After nailing twenty more targets in a variety of locations—rooftops, alleys, second story windows, doorways—he felt he had mastered this exercise. Every target had gone down fast.

"Halt."

Lights came on. The instructor strolled toward them. "Better." He gave a half smile. "Both of you did okay for this part." He studied them for a moment. "We'll now move the rest of your training to the deep-woods section."

Matt and Ralph turned and moved toward another area of the training facility. Matt thought this was like basic training in the Military and he and Ralph were reacting accordingly, but that was okay.

After they moved to the deep-woods section of the massive training facility, Matt and Ralph completed two grueling hours of additional practice.

Matt felt they had performed well.

During the wrap-up the instructor told them they should now be good enough to cope with a wide range of problems.

Matt shared the instructor's opinion and noted Ralph also appeared confident. He felt grateful everyone else they worked with was going to go through the same training he and Ralph had just completed. He knew that it was a necessary safety precaution.

As they left the facility, Ralph glanced at Matt. "We got a lot more training than I thought we would on how to spot snipers and what precautions to take."

Matt nodded slowly. "Yep, we got a lot of training in a lot of different environments . We should be as ready as we can be. To cope with a lot of danger."

Ralph gave a thumbs-up.

Chapter 34

In Moscow at Ivan's large mansion, Leon scanned the faces of the five men around him. Jake Bolton had joined the Belzex group for this meeting.

"It's time to solidify our plans for our new hi-tech datacenter." Leon looked at Chen. "Do you think you've synced up enough with Ivan, Bolivar, and Jorge?"

Chen nodded.

Leon looked at Bolivar, Ivan, and Jorge.

They also nodded.

"We've discussed a lot of possibilities and planned the implementation for a lot of advanced hi-tech products." Ivan looked back at Leon. "All based on the guidelines we agreed to earlier."

Chen nodded. "I think we're as synced up as we can be. We're ready to make some final decisions on the specifics of our hi-tech datacenter, structure wise and operations wise."

Leon narrowed his eyes and looked at Ivan. "Have you carefully checked your security in and around your mansion."

"We're well secured inside and outside," Ivan said. "My AI security system is on and fully functional." He looked directly at Leon. "Most of my best friends are KGB and they watch my back at the Kremlin and keep me informed on anything pertaining to my safety. That gives me and this location an extra level of security."

Leon nodded and glanced at his notes. "Okay let's keep moving forward." He scanned the faces around him. "As I've mentioned before, our new hi-tech datacenter needs to be fully functional as soon as possible."

"Do you have any additions to the capability we've agreed to?" Chen asked.

"No," Leon said. "Everything needs to be according to the plan we've specified."

Chen nodded.

"Is this new center going to be in a fortress like the ones you built before, and the one Hugo had?" Ivan asked.

"Yes," Leon said as he scanned the faces around him. He went on to highlight some details.

"Have you decided where we're going to build this new hi-tech datacenter?" Jorge asked.

"I think somewhere in Germany will be the best location," Leon said. "We can discuss it if you would like."

Everyone opted for a discussion.

After a discussion for a little over an hour, they decided on a location in Germany as the best site.

Leon looked at Jake. "You were in Hugo's fortress a good bit. Do you have any final comments on our selection?"

While keeping his gaze on Leon, Jake said, "I know you're planning to do some things differently than Hugo like you did last time, which puts a lot of things in a different light, but I think you made an excellent choice on the new site. Now we just need to get the new hi-tech datacenter built as quickly as possible."

"Okay, let's keep moving," Leon said. "I want to be able to start construction on the new site as soon as possible."

Leon spent the next two hours discussing the details he wanted with the other five men in the meeting. Jake Bolton contributed heavily to the discussion by providing a lot of details about Hugo's thinking when he built his hi-tech datacenter.

Toward the end of the discussion, Leon said, "There are a couple of things I still want to do differently than Hugo. I want a rectangular control room with walls thirty feet wide."

"All like you had last time, right?" Chen asked.

"Yes," Leon said, "with maybe a couple of little changes here and there."

Everyone kept their gaze on Leon and waited.

"I want one large flat screen covering the entire front wall with the capability to display thousands of various sized images from floor to ceiling."

Leon answered questions for thirty more minutes and then reached a logical stopping point.

"I guess we're in agreement enough to get started," Leon said. "Let's plan to begin construction as soon as we can."

"If we use the best of everything, technology, materials, etc., this is going to be super expensive immediately," Bolivar said. "We need to get the extra land and start the new crops of poppies right away. We'll need that additional stream of money real soon."

"I'll take care of that," Leon said. "Baker and Gibson won't be around much longer. I guarantee it."

"Do you have any new specific plans for getting them out of the picture?" Ivan asked.

Leon looked at Bolivar. " How many snipers do you have available right now?"

"Twenty," Bolivar said. "But some of them need more training."

"How many?" Leon asked.

Bolivar leaned forward. "Need more training?"

"Of course," Leon said impatiently. "What else would I be referring to?"

Bolivar shrugged. He hesitated for a few moments and then leaned toward Leon. "About ten could use some more training before I would consider them reliable."

Leon scowled. "Half of them?"

Bolivar nodded slowly.

Leon pressed his lips together and scanned the faces around him. "How about the rest of you. Does anyone have some super good assassins?"

Jorge, Chen, and Ivan all nodded.

"Are any of them talented snipers?" Leon asked.

No one commented.

"Well tell me something," Leon said.

"None of my assassins have ever needed to be a sniper," Ivan said. "They would probably be good, but I have no idea how good."

Leon looked at Jorge.

Jorge shrugged. "I can say exactly the same thing as Ivan did."

Leon pressed his lips together and then looked at Chen.

"I don't use snipers," Chen said. "I use my giants to kill people with their hands or I ruin them financially." He grinned. "I'm very good at both, and I've had plenty of experience in both arenas."

Leon nodded slowly.

After a brief silence, Leon narrowed his eyes and looked back at Bolivar

"Start putting your snipers through some more training immediately," Leon said. "I want some I can depend on within one more week."

Bolivar nodded. "You can plan on that being done."

Chapter 35

Late the next morning in Jacksonville, Ben Fulton sat at the large table in the main conference room in the FBI headquarters.

Kyle Rankin sat at the table with him.

Ben scanned the faces of the other people who sat around him.

Ralph Gibson, Matt Baker, Judy Caldwell, Todd Green, and Steve Baxter were there from Tuxtun.

Justin Mason, the Jacksonville Police Chief, and Vince Simmons, the Saint Augustine Police Chief were there.

Dan Maxwell (FBI) and Wendell Jones (CIA) were there from Washington DC.

"Paul contacted me last night. He had received some new data from NSA. I thought we needed more discussion about what's going on with Belzex," Ben said. "Kyle has the expertise in the subject matter. He can get us started."

Kyle shot a quick glance at Todd and Matt as he leaned forward and scanned the rest of the faces. "I discussed a lot of the information with Todd and Matt earlier and we all agreed to ask Ben to get everybody back together for another discussion. Luckily Dan and Wendell were still in town."

Everyone remained silent. Obviously waiting for Kyle to continue.

"There's a lot going on," Kyle said. "Belzex is starting a large construction project to build a new hi-tech datacenter in Germany." He paused. "NSA intercepted five key telephone conversations yesterday and decrypted all of them. We have a lot of new information on several different subjects to discuss. And they all involve Belzex, although the Hidden Empire was mentioned a few times."

Vince narrowed his eyes. "Were all the telephone conversations heavily encrypted?"

"Yes," Kyle said, "but NSA has a lot of updated technology. They can decrypt most phone conversation now, and they managed to decrypt all of these." He paused. "However, the use of technology is a leapfrog process. The parties engaged in various technological activities flipflop in having the upper hand. Right now, NSA does." He grinned. "And we're benefitting from it. NSA has just gathered a lot of good information."

"Do we know where they're building in Germany?" Wendell asked.

"Not yet, but we know they're building a hi-tech datacenter along with an accompanying power plant to take care of all the energy needs. It looks like they're going to need a lot of land," Kyle said. "They're planning on constant global surveillance and all that stuff." He paused. "Basically, the same stuff they were doing before, but their capability is probably better now. The related technology is improving fast."

Wendall gave Kyle a stern look. "Did the phone conversations reveal if Leon is going to try to conquer the world like he did before and like Hugo did?"

Kyle grunted. "Yes he is and he's setting everything up to be able do it."

"Remember Leon and Hugo are both descendants of Hitler

and they probably have the same genes," Matt said. "The urge to conquer the world is probably irresistible."

Ben glanced at Matt and Ralph. "It's going to cost a lot of money to build this new hi-tech datacenter and to make sure they have the power they need to operate it in the manner they want to. They're probably going to need more money than they have immediately available."

Matt and Ralph both nodded.

"That makes increasing their money flow from their illegal drug trafficking even more important," Ben said. "I think Leon is feeling even more pressured to get rid of both of you. He obviously thinks he has a better chance of getting Tuxtun's land if you're out of the picture."

Ralph gave a slight shrug. "No argument there, but I think we're as prepared as we can be. We have secure parking, so we don't have a threat from snipers while we're going to and from our cars at home and at work, and we keep all our security systems at work and home updated with the latest technology."

Ben nodded. "But you still need to be careful when you're driving between home and work. Leon probably will have access to some world-class snipers who are good at picking people off while traveling in their cars. We can be certain that Belzex has a good supply of them."

"Ralph and I have discussed that a lot," Matt said. "We have heavily tinted windows with bulletproof glass, and we take different routes to work every day."

"Once Leon gets his new hi-tech center fully operational that won't be any big obstacle for him," Ben said, "He'll use AI to map out all your routes and look for patterns. Plus, he might have enough snipers to cover every possible route."

Matt nodded. "We've thought about him looking for patterns. We try to be unpredictable. We bounce around among the routes randomly. We don't use any set system to plan our routes, and we pick our own individual routes at random while remembering to mix them up as much as possible."

"I'm sure Leon will think about that too," Ben said.

"Remember, he'll have the best in AI to predict your route, and he'll use multiple snipers so he can cover a lot of different ones, maybe all of them." He looked at Dan. "I know you're an expert on this very subject. Give us some advice."

"I've talked about this before, but it's worth repeating," Dan said. "There are some mathematical stochastic applications that can produce some good predictions in situations like we're talking about." He paused and scanned the faces around him. "One of the specific processes is called Random Walk. The classical explanation is that of predicting where a drunk leaving a bar at three o:clock in the morning will be after a few hours. It's often used in law enforcement to help predict outcomes from random activities." He made a few more comments and then his glance centered on Ralph and Matt. "I wish I had some more relevant things to tell you. I'm sorry but I don't have any additional advice that would of practical use to give you right now."

Matt and Ralph both nodded.

Ben leaned forward. "I know you're going to stay alert and be as careful as you can. That's my main advice" He paused. "You might consider wearing tactical helmets and bulletproof vests."

"Since we have bulletproof windows, I think that will take care of that concern." Ralph looked directly at Ben.

Ben nodded and didn't comment.

"All Ralph and I can do right now is play defense the best we can," Matt said. "And that's what we'll do."

Ralph gave a thumbs-up.

"We'll give you the best support we can," Ben said. "I'll ask Paul to keep NSA informed. Maybe they can do a little extra to try to spot a sniper in the target area with one of their surveillance drones."

"I'll ask Joe to do the same on the CIA side." Wendell gave a slight shrug. "It's the least we can do. I'll also ask Joe to coordinate with Paul and keep in close touch with NSA on emphasizing the importance of this."

There was a brief silence.

"When you bring Leon's possible use of snipers into the picture it's a whole new ballgame." Dan looked at Ralph and Matt. "When you go shopping or something where you get out of your car, there are a lot of different strategies a sniper might use, especially if they're very good and can pick people off at distances of over four hundred yards."

Ralph and Matt nodded.

"We'll be as careful as we can," Matt said. "We're already doing everything practical we can do."

Another brief silence ensued.

Ben looked at Justin and Vince. "Do either of you have anything to suggest?"

"Not really," Justin said. He looked at Matt and Ralph. "What you told us you're planning on doing sounds good to me. I don't have anything to add to it."

Matt and Ralph nodded and then tuned toward Vince.

"I'm thinking hard, and I don't have anything extra to suggest," Vince said. "You're already doing, or planning to do, everything I would recommend. You'll just need to constantly stay alert and use good judgement." He grinned. "And I don't think either of you has any problem with that."

Everybody laughed.

"Well, I guess we're done with that subject for now," Ben said. He looked at Kyle. "Do you have anything else you recommend we discuss in this meeting?"

Kyle nodded and the meeting continued for another hour.

Chapter 36

The next day at Leon's mansion in Germany, Leon scanned the faces of the five men around him.

As was usual lately, Jake Bolton had also joined the group.

"We need some more discussion about the operation of the joint financial office we're going to establish in Jacksonville," Leon said. "We still have a few loose ends we need to tie together."

"I want to better understand the purpose of this office?" Chen asked. "It hasn't worked before, and I don't quite get the benefit of doing this again."

Leon paused for several seconds. "I think the best way to describe the purpose is that having this office will allow us to stay in close touch with all the global financial institutions without having to make a lot of phone calls." He narrowed his eyes. "Phone calls can be intercepted by NSA, the Hidden Empire, and others."

"You told us that before," Chen said. "And it didn't work

like we expected."

Leon narrowed his eyes. "We have more capability now. We can defend ourselves better."

"Can't we encrypt all the calls so they can't be intercepted by NSA or by the Hidden Empire?" Chen asked.

"You have an advanced AI company with even more advanced technology than you've had before, you tell me," Leon said.

Chen tightened his lips and hesitated for a few moments. Then he looked at Leon. "We do have more advanced technology than we've had before. We might have the capability to do that." He paused. "I'm willing to give it a try."

Leon nodded. "Good. Go right ahead and keep me posted on your progress."

Chen gave a thumbs-up.

"Being able to make all the encryption work might be temporary," Ivan said. "There will always be a lot of experts in NSA and the Hidden Empire working full time to break the encryption."

Chen didn't comment.

Leon figured Chen knew Ivan was right.

"We'll give it another try anyway," Leon said. "Now let's do some planning on setting up the office. We don't have any time to waste."

Ivan looked at Leon. "So, this time, who do you want to occupy this office?"

"Same as before," Leon said. "We want representatives from all the major global financial organizations, including the Federal Reserve. That's a major reason for us to locate it in a convenient location in the United States. We can meet with all the necessary Federal Reserve people and discuss a lot of issues without having to make phone calls."

Bolivar removed an iPad from his briefcase. "Give me the names of the financial institutions again. I'm not an expert in finance and I want to do some more research. I know we're getting a large cashflow from the shell companies you've established and we're going to start getting a lot more. I want to understand the big picture better."

Leon nodded. "Again, we need elements of the World Bank, the IMF, the WTO, The Federal Reserve, and what used to be the International Acceptance Bank of New York in this joint office to facilitate and closely coordinate all our financial activities." Leon paused and looked directly at Bolivar. "We must coordinate all the activities of our dummy corporations with key players in the global monetary market. There are a lot of details. I'll tell you more about some of that later."

Bolivar glanced at his iPad. "Before you continue, give me the full names of the IMF and the WTO. I don't know much about the big picture in global finance."

The IMF is the International Monetary Fund, and the WTO is the World Trade Organization," Leon said. "Also, the World Bank has five different divisions. I'll explain more about how they fit into this in a few minutes."

Bolivar made some more notes and remained silent.

Leon scanned the people around him. Others were taking notes too.

After a brief pause, Leon continued. "The five branches of the World Bank are the International Bank for Reconstruction and Development (IBRD), the International Development Association (IDA), the International Finance Corporation (IFC), the Multilateral Investment Guarantee Agency (MIGA), and the International Centre for Settlement of Investment Disputes (ICSID)." He proceeded to give more details about how they functioned.

"That's a lot for me to digest," Bolivar said.

"Right now, you just need to continue to send the revenue you receive from your illegal drug trafficking to the shell companies." Leon paused. "That okay with you?"

"Of course," Bolivar said. He shrugged. "I just need to keep doing what I have been doing."

Leon looked directly at Bolivar. "I glad you're doing some research. It won't hurt you to know a little more about what's going on globally. It might occasionally help you in making some key decisions about your drug trafficking. Various aspects of all the global monetary organizations will be used by us to secretly funnel money to people we need to pay to facilitate our flow of drugs. Having an office with specialists from each of the major global monetary organizations makes things easier in a lot of ways when we want to concentrate on a specific area."

Bolivar nodded his head slowly and didn't comment.

Leon took a deep breath and leaned forward. "The elements of the WTO, the IMF, and the World Bank in this joint office will also be working closely with the Federal Reserve and they'll have a lot of face-to-face meetings to avoid their phone calls being intercepted." He paused." Our having the office in the United States also makes things a lot easier for our insiders at the Federal Reserve to maintain their cover."

"How so?" Jorge asked.

"Our person in charge of the office will find many valid reasons to call a lot of meetings where he requests representatives from the Federal Reserve attend," Leon said. "The meetings will be often enough where the leaders of the Federal Reserve will be happy to find volunteers to attend. Many of our insiders will volunteer occasionally. To avoid suspicion, they won't volunteer often, but I'm guessing the leaders will start asking them to go often because it's probably going to entail work no one wants to do."

Jorge nodded.

Leon laughed. "Our insiders will be going often and there won't be an ounce of suspicion attached to it. We'll be avoiding a lot of phone calls and that's great since NSA has improved their technology and phone calls can now be intercepted and decrypted no matter how good the encryption is."

"I can see where keeping the Federal Reserve closely involved in what we're doing helps in many ways," Ivan said. "From all the discussions we'll have about various details, it will make it a lot easier for us to make the right decisions about the money flow from our various shell companies."

A brief silence ensued.

"I'm still having trouble absorbing all the details of all the financial activities we have going on. Again, exactly where are the Channel Islands?" Bolivar asked.

"The Channel Islands are the islands of Jersey and Guernsey off the coast of England, located in the English Channel," Leon said. "There are some smaller islands in the mix, but they're considered a part of the bigger islands. They're all closer to France than Great Britain. They're thirteen miles from the northern coast of France and 85 miles from the southern coast of Great Britain."

"And some of the islands like Jersey are where they raise cattle, right?" Ivan asked.

"Right," Leon said. "These islands have provided a variety of tax havens, immune to inquiries from many global authorities, but remember they're not immune to the CIA. We'll have to be really careful as we go forward."

"And again, where's the Isle of Man?" Bolivar asked. "I've heard of it many times, but I still can't remember precisely the exact location."

"North of the Channel Islands, it has a central location in the Irish Sea between Northern Ireland and Great Britain," Leon said. "The financial community operates, for the most part, the same as it does in the Channel Islands."

Bolivar nodded.

There was a brief silence.

Leon scanned the faces around him. "Okay, I think we have enough details on the broader financial situation. Now let's concentrate on identifying anything else that we think will have a negative effect on Belzex."

No one made an immediate comment.

Leon paused and scanned the faces around the room.

"I'm still concerned about the long-range impact of the Hidden Empire," Jorge said. "Right now, I know they like the disruption we're causing, but I doubt they want us to get as powerful as we're planning on getting."

Jake nodded. "They'll definitely have a problem when our advanced technology gets close to being as powerful as theirs is." He paused. "I'll probably get a visit from someone soon." He looked at Leon. "I'm positive I'll be asked to give you a warning."

"Have you been able to discuss any specifics with your visitor in the past?" Leon asked.

"I've never really tried to," Jake said.

"Go ahead and try when you get a visitor this next time." Leon kept his gaze on Jake. "I want to try to determine how much leeway we might have in various areas."

Jake narrowed his eyes. "Our activities spread across a wide spectrum. Can you be more specific, like what kind of questions do you want me to ask?"

"Concentrate on finding out how sensitive they are about our improving our capability in AI," Leon said. "We might even improve our AI to the point where it's better than theirs."

"I'm sure they'll have a problem with that," Jake said. "There's no way they'll allow us to have technology more powerful than theirs."

Leon nodded slowly. "I'm sure you're right." He looked at Ivan. "Do you think we can get any help from the KGB if we need it for defensive purposes."

"That's going to be tricky," Ivan said. "But I'll do some inquiries and see if I have any success."

Leon paused and again scanned the faces around him.

"Does anyone have any more questions relating to the joint office?" Leon asked.

No one commented.

"So, everyone agrees with our plan?" Leon asked.

"It looks like you planned this out well," Jorge said. "Not having two hi-tech datacenters this time is a good move. Since the Hidden Empire will know all the locations we have, having a backup datacenter won't do us any good."

Everyone nodded.

"If Belzex operates effectively, there's going to be more

than enough money for everyone," Ivan said. "We'll all get richer, a lot richer."

"And if we work things out properly with the Hidden Empire, we should have a lot more power too," Chen said. "We should have enough power to make conquering the world a definite reality."

"We indeed have a lot of potential." Leon raised his right fist in the air. "I'm anxious to finalize all the details and get our new hi-tech datacenter built." He scanned the faces around him. "We all want more power and more money. I saw the potential benefits of getting us all together and here we are. We'll finalize the details real soon and get the new hi-tech datacenter built. We'll make sure we have it working properly as quickly as possible."

Everyone raised his right hand high in the air.

Chapter 37

The next day at Tuxtun in Jacksonville, Ralph sat at the round conference table in his office with four other people. Matt, Judy, Todd, and Steve filled the other chairs.

Ralph looked at Judy. "You're our financial expert, anything new on the financial picture at Belzex?"

"There's a huge amount of money flowing through their shell companies," Judy said. "Todd and I have been tracking it with help from Kyle, Dan, and Wendell." She paused. "And they, of course, are getting help from NSA."

"How much is the huge amount, as you put it?" Matt asked.

Judy grinned. "Over 200 billion dollars. We've been seeing about 10 billion in the same cashflow channels. That's a huge increase in just 24 hours."

Ralph, Matt, and Steve nodded.

"You know I'm a firm believer in following the money," Judy said. "All that money was funneled through shell corporations yesterday."

"How much over two hundred billion dollars?" Ralph

asked.

"Just a little over, two hundred billion plus another three hundred thousand to be exact," Judy said. "It was all paid to people and organizations in Germany."

"Is that where Belzex is starting work on their new hi-tech datacenter," Matt asked.

"It surely looks like it," Judy said. "They're buying a lot of land and paying a lot of people off. We're trying to get all the details as quickly as we can."

"Could you tell if any expenditures have gone toward setting up their joint financial office in Jacksonville?" Steve asked.

"It's not clear yet, but it looks like some of it went somewhere besides Germany. It's possible some might have made it to the United States and gone toward that," Judy said.

Ralph looked at Judy. "What about the Hidden Empire? Any more information about them?"

Judy shook her head. "Their transactions are very difficult to spot. They're much better than Belzex. All their transactions disappear when we try to track them. They seem to blend in with everything else." She paused. "Todd and I will continue our research immediately after this meeting ; the process the Hidden Empire uses is ultra complex. They obviously want to keep the real sources of all their funding hidden, and they're very good at that."

"We already know there's a very secretive global group of power-elites who are directly involved in the money flow for the Hidden Empire," Matt said. "We know they essentially control the major economic activities in the world. You might find some more useful data about the Hidden Empire as you continue investigating Belzex."

Judy nodded.

"I'm sure Belzex doesn't like our investigating their activity," Ralph said. "We need to stay prepared for any attacks from them."

A brief silence ensued.

Judy narrowed her eyes. "On the subject of the joint financial office, since Leon Schulz and Hugo Wagner each established a similar one before, it's probably for the same reasons."

"Along with all the other investigations we're involved in, Judy and I have uncovered some more interesting details about this joint financial office," Todd said. "There's nothing conclusive yet, and I don't want to get into all that now. But I just wanted to say we'll both continue to work hard to find all the answers we're going to need as we go forward with our investigation of all this financial activity."

Matt looked at Todd. "Does that include learning more about the Specter Matrix?"

"Hopefully," Todd said.

Matt nodded.

There was a brief silence.

"A couple more thoughts about the joint financial office being set up here in Jacksonville," Judy said. "Our Federal Reserve seems to be more involved in the early activity than we originally thought. I'm sure they intend to have a lot of face-to-face meetings to avoid their phone calls being intercepted by the National Security Agency. Having an office in the United States also provides more security for certain individuals working with the various global financial organizations." She went on to explain.

Matt nodded. "We know the NSA has improved their

technology. They can now intercept phone calls no matter how good the encryption on the call is."

"Since we think the activities of this new office are directly connected to the activities of Leon Schulz and Belzex, this could get very complicated for us," Todd said. "We know Belzex is going to use advanced AI for all their activities which likely would support an attempt for world conquest."

Steve grunted. "Yeah, and I'm thinking again about Leon Schulz being a descendant of Hitler. We know what he did last time and we know what Hugo Wagner did. I think we know what to expect this time around."

"Leon might end up having even better capability than Hugo did," Judy said. "I think we can be certain Belzex will soon have even better AI capability than they had before, and they'll apply it to accomplishing their goals."

"If they have ambitions of being dominant on the world stage, and if they have enough money to afford a super technological setup with the best artificial intelligence, spy satellites, and such,' Todd said. "They're going to be a real threat to all of us here at Tuxtun."

"They're for sure going to need a lot more income," Ralph said. "And that's going to make them more desperate than ever to get our land."

Matt nodded. "This situation will probably continue to get more complex and more difficult to untangle. We'll need to stay on our toes constantly."

Todd narrowed his eyes. "Too bad our Federal Reserve is involved in this. The owners of the Federal Reserve have always been masters of disguise as it applies to what they're really doing. They have a long history of covering things up."

"We already know what you're probably referring to, so

don't bother giving us another history lesson," Ralph said.

Todd shrugged. "Just wanted to remind everyone about the depth of what we're dealing with."

"With the exception of the Hidden Empire, I think we're all well aware of what we're dealing with and who we're dealing with," Ralph said.

"Regarding the Federal Reserve, many people connected directly to it have been ruthless about things over the years and their descendants are probably still that way." Matt paused and scanned the faces around him. "And we know many descendants of the founders are still in place."

Judy nodded. "Another example of genes being a dominating force."

Matt leaned forward. "Yeah, and they might be directly connected to a lot of the activities that this new joint financial office in Jacksonville is going to be involved in. That alone could lead to a lot more danger we need to deal with."

"Yep and having the Hidden Empire lurking in the shadows adds a whole new dimension to the situation," Ralph said. "Let's hope we can learn more about them and get a lot of things figured out real soon and then, if we need to do something, be capable of taking the necessary actions."

Steve scanned the faces around him. "Speaking of taking the necessary actions, I suggest we each start wearing a bulletproof vest again at all times during the day, even in the office. I think the danger is that great and it's smart to be as protected as we can. We're all wearing our Glock19 pistols all day, but I think we should take this extra step."

Todd nodded. "Belzex contains an AI development company along with two manufacturing companies, and we already know they can override our security system and break into our office."

"I've gotten something done about that," Steve said. "I got a lot of help from some of the best technology experts at the CIA. That problem is at least temporarily fixed." He went on to give some details.

There was a brief silence.

Ralph looked at Steve. "We bought some protective vests before. Do you still recommend the same type?"

"There are still two main types I recommend," Steve said. "One is made of amorphous silicon carbide. Another is made from a tungsten disulfide-based nanocomposite material, commonly referenced as lightweight nano-armor. Each one is over five times stronger than steel. Everything considered, I still recommend the nano-armor."

"So, we'll keep what we have, right?" Ralph asked.

Steve nodded. "We just need to start wearing them again as I suggested."

Everyone gave a thumbs-up.

Ralph studied some notes he had for a few moments.

"Okay," Ralph said. "I want to cover a few more things before we wrap up this meeting." He looked at Judy. "Since the international finance activities play a large part in most of the threats we're concerned about. I'm sure you have some more to say to enlighten us about certain matters."

"Yeah, I've been doing some more thinking, and I've come to a couple more conclusions that I want to mention," Judy said. She scanned the faces around her. "I've mentioned the big money flow increase." She leaned forward. "One aspect of that is the banks, that do what the International Acceptance Bank of New York used to do, also had a huge number of transactions yesterday. I'm pretty sure now that a lot of money has just been delivered to the right people to facilitate the planning and construction of the new hi-tech datacenter, the powerplant to support the extensive use of AI, the acquisition of the land for them, and the opening of the joint financial office in Jacksonville."

"They're moving fast, "Matt said. He turned toward Judy. "Like you've mentioned, they obviously need a lot of money for what they're doing." He looked at Ralph. "Leon will probably try again to get us out of the way, and real soon. It's obvious he's going to need the extra money flow from illegal drug trafficking in the not-too-distant future."

Ralph nodded. "Fireworks are going to start going off in all directions. At least we're trying to protect ourselves to the best of our ability. We'll start wearing our protective vests again immediately."

Chapter 38

That afternoon in New York City, Zeus sat with Odin and Jupiter at a round conference table in a secluded room at a secret location.

The door was closed, and the room had been swept for all types of electronic bugs.

"We need to continue to keep a close eye on what's going on with Belzex," Zeus said. "I don't like it. It looks like they're going to be some big trouble for us again. I don't think Leon has learned any lessons from what he went through before."

"It's Hitler's genes," Jupiter said. "You can't fight them. The urge to conquer the world is too strong."

Odin nodded. "Belzex is moving forward quickly with their global business plans. I think they're going to want a lot more world power soon, and they know that we're sitting out here as a big obstacle blocking them from what's obviously their ultimate goal."

Jupiter glanced at Odin and Zeus. "We still have people in place who have been monitoring their phone conversations constantly, right?"

"Yes we do," Odin said. "And we still have access to many

of their other conversations too. We're fully equipped with all the latest eavesdropping capabilities. We not only have our own advanced technology team, but we have insiders in all the big hi-tech organizations who have direct access to all the latest technology for these types of things. They put a lot of time and effort into keeping us well equipped. Everything is still working perfectly for us."

Zeus laughed. "Yeah, we wouldn't be masters of stealth and titans of influence without that support."

Odin and Jupiter also laughed.

"Well anyway," Odin said, "we know they're planning to build a new hi-tech datacenter in Germany and a powerplant nearby to supply all the electricity it needs."

Jupiter nodded. "And with their new expertise and capability, they'll probably have them up and running soon."

Zeus narrowed his eyes. "Yep, it's becoming more obvious that Leon Schulz has an uncontrollable urge for power since he's a descendant of Hitler, and he likely has the same plans as Hugo Wagner did. He knows he needs to destroy us to be able to achieve his goal of conquering the world."

"Yes and we'll obviously have the same issues and problems to deal with as we had with Hugo and as we had with Leon before," Jupiter said. "And we should certainly expect to have these same issues and problems and be fully prepared to deal with them as soon as they develop."

Zeus narrowed his eyes. "It looks like those issues and problems are going to be developing real soon. We know Leon's trying to eliminate Ralph Gibson and Matt Baker. I suspect as soon as he's done that, he'll be coming after us."

Odin took a deep breath and exhaled sharply. "Yeah they'll come after us real soon. We can expect it and stay prepared to deal with it."

"Is there anything specific you're worried about?" Zeus asked.

"Ther's a huge spectrum of things," Odin said. "We had better be prepared for all of them. If we wait very long to act,

Leon and Belzex my become more powerful than we are."

Zeus and Jupiter both nodded.

"We had better alert the KGB and make sure they're prepared to act when we need them," Zeus said. He glanced at Odin and Jupiter. "I'll plan to take care of that personally."

A brief silence ensued.

Zeus gestured to Odin. "Go ahead and name a few problems you're thinking about."

"We know Leon is again planning to establish Belzex as a large global conglomerate. He'll again set it up to be a large provider of food, legal drugs, fuel oil, and many other products," Odin said. "There are a lot of other conglomerates doing similar things, so they have good cover globally by being in a large crowd."

Jupiter rubbed his chin. "I still think Belzex is going to concentrate on trafficking illegal drugs. There's a lot of money in that and they're going to need it to fuel their massive AI activities. They're going to promote legal drugs like Morphine and Codeine, but they're going to concentrate more on producing Heroin and Opium. They'll make a lot more money that way, and as we know, they're going to need a lot more money to properly maintain the gigantic AI operation they're setting up in Germany."

Odin nodded. "Belzex will use the legal part as cover, but they'll be putting most of their effort into the illegal part and setting up their global cashflow to balloon into an astronomical amount. That will help them pay for all the things they'll need to use to increase their power." He laughed. "And I mean their power in every way, personal power, and electrical power."

Zeus and Jupiter laughed too.

In a few moments, Zeus said, "This is all serious stuff, but it helps to maintain a sense of humor, so it doesn't get you down." He paused. "From listening to their phone conversations, it's obvious they know all about increasing their power in every way." He studied Jupiter and Odin for a few moments. "We have a big problem that's going to be coming up. We should never have let Leon off the hook. We should have destroyed him when we had a chance."

"But we all thought he had learned his lesson and could again become a major distraction to keep law enforcement busy with the things he was doing," Odin said. "That would be a benefit to us. However, he's well on his way to crossing the line again. So, if he continues in the direction he's going, we'll have to eliminate him and his whole organization."

Zeus nodded. "We must prepare to eliminate Belzex before they get too powerful. They're obviously doing all the things they need to do to get to where they want to be."

Jupiter leaned forward and shook his head slowly. "We certainly know firsthand an enormous amount of money can be a big help in gaining an enormous amount of power."

"We'll continue to keep a close eye on what they're doing and take the necessary action as soon as we need to," Zeus said. "Also, we'll make a special effort to keep our technology updated. We can't let them get ahead of us."

Chapter 39

The next morning at Leon's mansion in Germany, Leon, Jake, and Bolivar sat in comfortable chairs in a spacious den area.

"Hugo had a difficult time trying to eliminate Baker and Gibson," Jake said, "even though he had all of his hi-tech working." He looked directly at Leon. "You've had the same difficulty too."

Leon narrowed his eyes and leaned toward Jake. "What advice would you give us?"

Jake shrugged. "Nothing Hugo did to eliminate them worked. He even used his hi-tech for them to enter highly secure areas so they could catch Baker and Gibson by surprise. He did that several times, while they were at work at Tuxtun and at each of their homes." He paused. "In your original version of Belzex, you tried all the same things, and you didn't eliminate them either."

Leon nodded. "Yeah, nothing has worked so far, but we'll change that. We have a lot of experience from what we've done before, and we're going to have a lot more advanced technology very soon."

Jake shook his head slowly. "Nothing has worked any of those times you and Hugo tried. Baker and Gibson were too skilled and too resourceful." He paused. "And you can expect them to be that good again."

Bolivar stared at Jake "Surely you can think of something that can help us. I have the best assassins on the planet, and they all have a lot of experience." He continued to stare at Jake. "Any useful tips will be welcome."

Jake nodded and slowly rubbed his chin. He seemed to be thinking hard.

After several more minutes had expired, Jake regarded Leon carefully. "The only thing I can say is that I would advise you to further analyze Matt Baker's and Ralph Gibson's skills and develop a better plan than you had before to cope with them."

Leon nodded. "I'm for giving that a try. It's going to be a while before we get our new hi-tech datacenter running to our satisfaction." He looked at Bolivar. "Plan to send a couple of your top assassins. Caution them repeatedly about being overconfident. Make sure they find a way to practice what they're going to do." He paused for a few seconds. "Probably figuring out a way to confront Baker and Gibson early one morning on their way to work would give your assassins their best chance."

Glancing at Jake, Bolivar asked, "You advised us to develop a brilliant plan. Do you have anything specific to suggest?"

"Hugo had a lot of brilliant plans and none of them ever worked." Jake looked at Leon. """You had a lot of brilliant plans too, and none of them worked either. You and Hugo have tried everything." He paused again and leaned toward Leon. "My best suggestion is for you to work on it for a while, think about trying something entirely different, and see what you can come up with."

Leon held each arm to his side with his palms up. "You've

already mentioned that we've tried everything."

"Yeah but there are still some possibilities that could have a chance to work," Jake said. "Go ahead and have some more tactics discussions. I'll go check on some things and come back. I might have some more advise for you."

Leon and Bolivar nodded. Neither made a comment.

There was a brief silence.

Jake glanced at his watch and then looked at Leon. "If it's okay with you, I'm going to leave now. I'll check on some things and be back with you in a couple of hours."

Leon gestured with his right hand and Jake left immediately.

Chapter 40

A couple of hours later, at Leon's mansion in Germany, Jake entered the lounge area and selected a chair to sit in.

Leon and Bolivar stopped talking and looked at him.

"Have you developed a plan yet?" Jake asked.

"We think so," Leon said. "We can see what you think about it, or we can listen to what you have to say about what you checked on."

"I'll listen to your plan first and then give you the new information," Jake said. "It might be good to merge some things together."

Leon gestured to Bolivar. "We're planning to use two of your best assassins. You go ahead."

Bolivar leaned forward. "We checked with Chen about all the AI devices that are working in the warehouse right now. We concentrated on the ones that will allow our assassins to get into the Tuxtun headquarters building undetected."

"NSA has probably mode a lot of improvements in their detection devices and Tuxtun probably has all the improvements," Jake said. "Does Chen think our assassins will still be invisible to all the surveillance devices at Tuxtun?"

"Chen's not sure," Bolivar said. "However, he emphasized that his highly skilled team has also made a lot of improvements. In fact, they installed a huge number of new features this morning. He thinks there's a good chance we can still get in undetected."

"So, you're still going to use the same plan as you did last time?" Jake asked.

"Not exactly the same," Bolivar said. "The assassins will be different, and I think they might be more skilled. They'll use a different route inside Tuxtun and some different tactics." He went on to explain.

Jake listened carefully and then looked at Leon. "That's not too shabby. It might work."

"Like I mentioned before, I'm willing to give it a try," Leon said. "We don't have a lot of other options right now since our new hi-tech datacenter isn't yet available."

There was a brief silence.

Leon leaned toward Jake. ""Tell us what you've been looking into."

"I was checking with Ivan about some of the things he and Jorge were working on with their manufacturing processes. We have a lot of good possibilities coming up," Jake said, "but none of them are ready yet." He went on to explain.

"So, you think the plan Bolivar and I have devised is the best we can do right now?" Leon asked.

Jake gave a short laugh. "Like you mentioned: It's worth giving it a try." He looked at Bolivar. "And the assassins you're going to use are Pablo Garcia and Rafa Lopez?"

Bolivar gave a thumbs-up. "And they're already in Jacksonville."

Chapter 41

The next day, Pablo and Rafa sat in some comfortable chairs at a small table in a secluded corner of a large hotel lobby.

The hotel was only two miles from the Tuxtun offices.

Pablo scanned the area again and saw no one else was in the area. He reached into his briefcase and removed a set of photographs, which he immediately spread out on the table.

"I'm glad Bolivar managed to get these for us," Rafa said. "Our mission would be a lot more difficult without them."

"We certainly can't complain about lack of support," Pablo said. He began sorting through the photographs. "Baker and Gibson still look to be in top condition."

Rafa nodded. "We'll need to be prepared to face a big challenge. We'll need to be sharp and make sure we'll be at the top of our game."

"Yep, just like we were told. We need to expect them to be as skilled as any combatants we've ever faced," Pablo said. He paused and stared at Rafa for a moment. "But this is nothing new for us, right?"

Rafa nodded. "Right, we've done this many times before

and we've always won. We've beaten some of the best gunmen in the world."

Both men continued sorting through the photographs on the table.

Each of the photographs had a note at the bottom of it.

Pablo read every note carefully and noticed that Rafa also did that, maybe even more carefully than he had.

Both men remained quiet for as they shuffled through the photographs.

"I think I can recognize either one of these guys instantly from any angle," Pablo said. He narrowed his eyes and looked directly at Rafa.

Rafa nodded. "Same here. And we'll need to react instantly when we recognize them. Our intelligence states they can draw their pistols and fire in a half second."

"Normally that would be a challenge, but we'll have an advantage. Their speed matches ours, except we'll have our pistols in our hands."

"That's not such a good idea. We'll attract too much attention walking around with drawn pistols."

Pablo stared at Rafa for a moment and then said, "You're right, some people will probably be walking around out in the hallways."

Rafa narrowed his eyes. "Possibly a lot of them."

Pablo grunted. "We don't want alarms going off immediately, so we'll need to keep our pistols holstered as we're walking past the offices and meeting rooms."

"Probably until we're ready to rush through the doorway of the room Baker and Gibson are in," Rafa said.

Pablo nodded. "We'll evaluate the conditions are and make that decision at the last second."

Rafa gave a thumbs-up.

A brief silence ensued.

"You're going to wear a coat, right?" Rafa asked

"Of course. Best way to keep your pistol hidden and have easy access to it." Pablo gave Rafa a long stare. "That's what we always do."

Rafa shrugged as he returned Pablo's stare. "I was just making sure we were thinking in sync. We can't make any mistakes of any kind." He paused. "We haven't worked together in over three years. Are you still using your Heckler & Koch USP?"

Pablo nodded. "And still with a 15-round magazine of 9x19mm Parabellum cartridges. I've never found it necessary to change. What about you?"

"I'm still using the SIG Sauer P226 with 9x19mm Parabellum cartridges,. I also have a15-round magazine," Rafa said. "Just like you, it's always worked well for me, and I've never found it necessary to change."

Pablo tightened his lips and narrowed his eyes. "Looks like we have good firepower. Let's hope we have some good luck too."

Rafa gave a short laugh. "That never hurts."

"Want to discuss the strategy we'll use when we walk through the front doors?" Rafa asked.

"Do you have some recommendations?"

"I don't think we should stay close together. It would be better for us if employees at Tuxtun don't know we're together."

"Should we go the same direction, or should we split up completely and go different directions?"

Rafa shuffled through the photos on the table in front of him and grabbed a piece of paper. "Ah, here it is, a floorplan of the building. We'll give it a quick look." He moved it closer to Pablo.

Both men studied it for a few moments.

"The main entrance, Gibson's office, Baker's office, and all the conference rooms are well marked," Pablo said. "We'll need to make some choices on our exact strategy about where to go after we get inside the entrance to the building."

"Going straight to Gibson's office seems good to me." Rafa narrowed his eyes and leaned toward Pablo. "That's where I'm guessing they'll be."

Pablo pressed his lips together for several moments. "No matter which way we go we'll need to be super alert and ready to react instantly." He looked back at the sheet of paper showing the floorplan. "So, do you think we should plan to go straight to Gibson's office first."

Rafa thought for a few seconds and then nodded. "You lead the way, and I'll trail about ten yards behind you. You can wait a few seconds outside the doorway for me to catch up before you go in."

"That sounds good," Pablo said. "We have a plan."

Rafa gave a thumbs-up. "We'll just have to stay alert and execute well when we find our targets."

Pablo flashed a big grin. "But we'll need to be careful when we're walking in the hallways to not be conspicuous."

"We can do that. We're experienced assassins."

Pablo nodded. "Yeah, we've ambushed good fighters plenty of times in a lot of different situations. We should have no problem doing it again in this situation."

Rafa reached for his secure satellite phone. "I'll call Bolivar and tell him we're ready to leave the hotel lobby."

Chapter 42

In Germany, Leon Schulz sat in his study in his elaborate home and looked at Jake and Bolivar.

"I checked with the person Hugo always used to evaluate assassins," Jake said. He paused for several moments. "I decided to find out what he had to say about Pablo and Rafa." He looked at Leon and shrugged. "I figured it might be a good thing to do."

"And?" Leon asked impatiently.

Jake narrowed his eyes. "Both Rafa and Pablo are near the top of the global list of assassins rated by skill level."

"So, they should be capable of getting the job done," Leon stated as he regarded Jake carefully.

Jake nodded slowly. "They're also the best we have available at this time."

Before anyone commented further, a stranger entered the study through the open doorway.

Leon narrowed his eyes and stood. "How did you get through our security?"

Jake and Bolivar also stood.

Leon knew the latest updates from Chen had been

installed. Chen had assured him his security system could not be penetrated.

The stranger laughed. "I'm with the Hidden Empire and we still have a higher level of AI than you do, which gives us more capability than you have." He paused. "That'll be good for you to keep in mind."

Leon, Jake, and Bolivar stared at the stranger and made no immediate comment.

"Our capability is such that we can make ourselves completely invisible to anything but human sight no matter how good the detection system is," the stranger said. "That's why no motion detectors or any of your other security devices were able to alert you." His grin expanded as he stared at Leon.

"Hugo had that capability, and we had it too," Leon said. "We now have it back again."

The stranger nodded. "We know."

Leon didn't comment.

"You don't have it as perfected as we do," the stranger said. "Remember, we're way ahead of you in every aspect of advanced technology."

The stranger looked directly at Leon. "We know everything you're doing, including your sending the two assassins to kill Ralph Gibson and Matt Baker."

"Eliminating them will cause a lot of disruption," Leon said. He stared at the stranger for a moment. "Jake tells us that you like disruption because it helps you to stay well hidden from all the advanced intelligence law enforcement uses."

Jake nodded.

The stranger shrugged. "Sometimes you do us some good as long as you don't cross the line."

"Since we're going to soon be causing a lot of disruption and doing you a lot of good, can you help with the AI bit?" Leon asked.

The stranger narrowed his eyes and looked at Leon. "What help do you want?"

"We would like to have the full capability you just used,"

Leon said. "We want to always be able to make our assassins invisible to everything but the human eye, even against very advanced detection systems."

"My cohorts and I have discussed that many times," the stranger said. "And we decided to stay out of what you're doing unless you cause us problems." He shrugged. "We don't really need all the extra disruption you'll cause. The way the world is right now, we have plenty of disruption anyway."

A brief silence ensued.

"Then why are you here?" Leon asked.

"It looks like you're about to cause us some problems. I'm here to caution you about some things, to give you an early warning, so to speak," The stranger gestured toward the table. "May I take a seat?"

Leon waved him over.

After each of the four men settled into his chair, the stranger looked at Leon and said, "You're planning to build a new hi-tech datacenter with even more capability than you've had before."

Leon nodded.

The stranger narrowed his eyes. "We're not going to allow you to have too much capability. If you're planning to probe what we're doing that will be crossing the line, and we'll eliminate you and Belzex." He paused. "We eliminated Belzex before. This time we'll also eliminate you."

"Why?" Leon extended both arms to the side with his palms up.

"Because it looks like you previously had a goal like Hugo Wagner had to conquer the world, and it looks like you might have that goal again. Last time we decided to let you live, but it looks like you didn't learn your lesson. "The stranger raised his eyebrows and looked directly at Leon. "We'll allow you to have some advanced technology, but don't try to use it on us."

Leon frowned but didn't comment. He noted the stranger continued to stare at him.

"This is the only warning we plan to give you," the stranger said. “If you don’t stay in your lane, you’ll suffer the consequences.”

Leon, Jake, and Bolivar all remained silent.

“Do you have any more questions?” the stranger asked.

“So, you’re okay with our killing Matt Baker and Ralph Gibson, right?” Leon asked.

The stranger grunted. “The point has already been made that it will cause a lot of disruption and we’re okay with that. More disruption, although we don’t need it, won’t hurt anything.” He gestured toward Jake.” And he correctly stated that we like disruption because it helps us maintain secrecy. I mentioned earlier that we don’t need a lot more disruption, but we’re okay with you killing Baker and Gibson. You might be doing us a favor and saving us some trouble. They’re trying to analyze the cashflow in our Specter Matrix, and they could end up causing us a problem. We might need to eliminate them too.”

“So where does that leave us?” Leon asked.

“We’ll let you go ahead with your plans,” the stranger said. “We just won’t help you succeed.”

There was a brief silence.

“In the grand scheme of things, a little more disruption certainly won’t hurt anything,” the stranger said, “so that’s why we’ll allow the killing.”

“So why not help us improve our AI capability to basically ensure our success?” Leon extended both hands to his side again with the palms up.

“I told you we’ve already discussed the situation and decided to stay out of it.” The stranger stared at Leon for several moments. “It can go either way as far as we’re concerned. You can succeed in killing them or you can fail. We don’t care.”

“Why don’t you care?” Leon asked.

The stranger narrowed his eyes. “We have our reasons and that’s all I need to tell you.”

After scanning the faces around the table, the stranger stood. “I’ve delivered the message, and you need to keep everything I told you in mind.”

Leon raised a hand. “Before you leave, I have one more question.”

The stranger motioned for him to go ahead.

“How were you able to achieve the level of AI capability that you obviously have?”

The stranger laughed. “You should know I’m not going to comment on that.” He looked directly at Leon. “You need to remember what happened before to your organization and to Hugo and proceed accordingly. We support what you’re doing now, but we won’t support your trying to get rid of us, or your trying to conquer the world.”

Leon nodded slowly.

After giving all three individuals a stern glance, the stranger turned and left.

Leon reached for his secure satellite phone. “I’ll check with Chen and let him know his updates to our security didn’t work well enough.”

Chapter 43

The next morning at Tuxtun in Jacksonville, Ralph sat at the round conference table in his office. It had seating for eight.

Ben, Matt, Judy, Todd, and Steve filled five of the other chairs.

"I have some new information from NSA to share with you," Ben said, "but before we get started, I want to check on something." He scanned the faces around him. "Are you wearing your protective vests?"

Everyone nodded.

"We can expect the assassins to be using powerful weapons with high caliber ammunition," Ben said. "Some bullets will be able to penetrate the vest to some extent but wearing it will provide additional protection."

Matt leaned toward Ben. "Are you wearing one?"

Ben nodded. "Same type you're wearing. I've worn it for several years. It's worked well many times and probably saved my life in every one of those instances."

"We know Leon wants to kill all of us as soon as possible, and I'm sure he plans to keep trying until he succeeds," Ralph said. We're trying to be as safe as we can, so we'll stay on our toes and use every precaution we can."

Ben continued to nod. "Yeah, you need to do everything you can to protect yourselves, especially when you're going up against worldclass assassins, which I'm sure Leon has a good supply of. They'll keep coming in various combinations."

There was a brief silence.

"There's a lot of new information from NSA because they've improved their technology. One of the more interesting items is that the Hidden Empire has again paid a visit to Leon and gave him a warning about what he was planning to do," Ben said.

"I know you told us NSA had improved their technology, but do you have any specifics on how they found that out?" Matt asked.

"Just like before, I asked some questions, but I didn't get any specific answers," Ben said. He shrugged. "Of course, I wasn't expecting any specific answers. They must protect their methods and not reveal too much of their capability, not even to the FBI."

Steve leaned toward Ben. "Did NSA give you any details on some other things?"

"Yeah," Ben said. "We have some more things to think about and that's why I came over to discuss them." He paused and scanned all the faces around the table. "NSA knows that the Hidden Empire is keeping a close watch on all of us and that the technology they're using is powerful enough to do that." He looked directly at Ralph. "I know those of you in this room are working hard to improve your AI here at Tuxtun." He glanced at Todd. "And I know you're leading that effort."

Todd nodded. "We're getting a lot of help from law enforcement, Kyle in particular, and we appreciate it."

Ben nodded. "And he's getting a lot of help from NSA. Too many powerful people with evil goals are getting too many big payoffs through the Specter Matrix, and we need to find a way to put a stop to that." He looked at Ralph. "We appreciate all of the effort from everyone here at Tuxtun."

“We’re all on the same team and we’re glad to help as much as we can,” Ralph said.

Ben looked at Ralph and Matt. You guys have been a big help to law enforcement in a lot of ways, and we appreciate it. We don’t want to lose you, and we’ll do everything we can to help you survive the threats you’re facing.”

Everyone gave a thumbs-up.

Ben mentioned a few more details and the discussion continued for over an hour.

Matt, Todd, and Judy were especially helpful in tying a lot of scatted facts together and bringing them into focus.

Ralph looked at Ben. “I don’t know about everyone else, but I would like to take a break for about thirty minutes and check on some business matters.” He grinned. “We still have a business to run.”

“Take your time,” Ben said. “A break would be good. I have some things to check on too.”

“I would also find a break helpful,” Judy said. “The global financial picture changes constantly and it’s good to try to keep up with any big changes. Also, Todd and I have some other things where we need to check the status.”

Chapter 44

In Germany at Leon's mansion, Leon looked at the other five men sitting around him in his luxurious den.

Leon checked his watch. "Well Pablo and Rafa should be about ready to eliminate Baker and Gibson."

"I'm keeping up with them and their location," Bolivar said. "They're in the right area and should be going into the Tuxtun headquarters building within the next five minutes."

Everyone nodded.

Leon scanned the faces around him. "Even though we're only building one hi-tech datacenter, the extra cash flow we'll get from using the extra land to grow poppies will still be needed to achieve our goals." He paused. "The tremendous capability of the AI we're going to use will require a huge amount of energy and that will entail tremendous costs. We don't want to end up with any cashflow problems. We need that extra land to produce the amounts of opium and heroin we'll need to produce the extra income."

Everyone nodded.

There was a brief silence.

Leon looked at Bolivar. "You're watching the progress of Pablo and Rafa?"

"I'm keeping up with every detail," Bolivar said. "Chen has provided a lot of new AI capabilities and I'm using some of the latest features."

"Where are they now?" Leon asked.

Bolivar looked down at the device he held in his hand. "They're moving in the direction of Tuxtun." He punched a few buttons. "They're exactly fifty yards away and moving at 5 miles per hour."

"So, they're about ready to go in the entrance," Leon said. "We don't have long to wait." He grinned. "It will feel really good to get rid of those two jerks at Tuxtun. Maybe whoever takes over the ownership will have enough sense to sell at ten million dollars an acre."

Bolivar nodded. "That will be quite an addition to their profits right away. I don't expect them to turn it down. The new owners will want to look good to all their shareholders."

"Yeah," Leon said. "That's pretty much a necessity for any normal corporate CEO."

Bolivar continued to scan his device.

"I'm glad our technical experts were able to develop and add that additional AI enhancement to the apps on Pablo's and Rafa's phones," Chen said. "Their being able to look at the floor plan and see the exact locations of Gibson and Baker should be very useful."

"Yeah," Bolivar said. "We wanted them to have some additional advantages wherever possible."

Bolivar looked at Leon. "In spite of no help from the Hidden Empire, Chen's team has also been able to develop the capability to make our assassins invisible to all the security monitors. Chen told me that NSA has made a lot of improvements to their security technology, but he thinks he now has our capability improved enough to keep that from working properly."

Chen nodded.

Leon turned toward Chen. "Are you sure about that?"

"About our having the capability to keep NSA's improved detection technology from working properly?" Chen asked.

"Yes," Leon said. He tightened his lips.

"Things are always changing. You can never be absolutely sure." Chen gave a slight shrug. "My technology team knows how important this is, and they've done more testing than usual. The results have been good."

Leon nodded slowly and didn't comment.

"We're always working hard to stay ahead of NSA and to at least stay up with the Hidden Empire," Chen said.

"Good." Leon narrowed his eyes. "As we go forward, there will probably be a lot more people we'll need to get rid of in various ways, and being able to successfully penetrate their security systems, no matter how advanced, won't just be useful, it will be absolutely essential."

"I understand that well," Chen said. "That's why I have my best people working on it in multiple shifts. We'll make sure our capability is where it needs to be and stays there."

"Are you constantly checking the details multiple times each day?" Leon asked.

Chen nodded.

Leon gave a thumbs-up.

Chapter 45

After the break, Ralph continued the meeting in his office and looked at Ben. “You mentioned you still had some new information to discuss. I guess we’re as ready as we’ll ever be to keep going.”

“That joint financial office is now open here,” Ben said. “They bought an older building and refurbished it. They’ve already moved in, and NSA has observed there’s a lot of activity in the building and on the grounds.”

“Any details about the activity?” Matt asked.

“The organizations we expected are involved,” Ben said. “The activity looks about the same as it did in the Arlington, Virginia office and when Hugo Wagner established his joint financial office in the Jacksonville area before. That’s no surprise since we’ve known from the beginning that the building is for the same purposes as Leon had before and as Hugo had.”

“Should we pay them a visit like we’ve done before? Ralph asked.

Ben gave a slight shrug. “We didn’t learn anything useful about their activity, and they had a chance to knock us off. We can be sure they'll be forewarned about our visit." He scanned the faces around him. "We know Leon has recovered a lot of his AI capability now, and I'm sure the results will be the same relative to our learning anything useful."

"Has NSA intercepted any more phone conversations about the status of the hi-tech datacenter Leon is in the process of building?" Matt asked.

Ben nodded. "They're intercepting a lot of calls. I don't know if we're getting all the content from the conversations because the encryption on the ones we intercept is better than before." He paused. "NSA keeps improving their technology and we've been able to decrypt most of what we've intercepted. We've learned a lot. However, we know we have a lot more to learn."

"Do we know when Belzex's next meeting is?" Ralph asked.

"According to the last report I received from NSA, it's going on right now in Germany," Ben said.

"Did you learn any more about the schedule for the assassins?" Matt asked.

"No," Ben said. "Any information about that was probably in some of the calls we haven't decrypted. But we did learn they have a way to shield the assassins from detection from our security monitors." He looked at Ralph and Matt. "They can get in and they can detect where you are. However, NSA has made some improvements and they're remotely installing them on your detection system. Right now, we'll know the assassins are here when they show up."

Matt narrowed his eyes and glanced at the open doorway. "We probably should keep the door closed and locked."

"Our security alarms will go off quickly," Steve said. "Since we'll know they're here, we can go after them or close and lock the doors then."

After a little more discussion, everyone agreed on the course of action. Since they would be forewarned immediately if any intruders entered the building, and since Ralph's office was a long way from the entrance, they would leave the door open for now.

Ralph took a deep breath and exhaled slowly. "Well, we know we're in a lot of danger and I guess we're as forewarned about the threats as possible."

Matt nodded. "If we stay alert, I think we can handle things."

"Remember these assassins will likely be worldclass," Ben said. He paused and leaned toward Ralph and Matt. "But you've done well before and I think you can do well again."

There was a brief silence.

Ralph looked at Ben. "Thanks for all the additional information. We'll use a lot of caution as we go forward."

"Good luck to all of us," Ben said.

Chapter 46

Rafa and Pablo walked toward the main entrance at Tuxtun.

The necessary security applications to allow them access into the building were loaded into a phone each assassin had in a coat pocket.

Each assassin had his favorite pistol in a holster positioned on his right hip under his coat. Pablo had his Heckler & Koch USP. Rafa had his SIG Sauer P226. Each pistol had a 15-round magazine of 9x19mm Parabellum cartridges.

Pablo glanced at Rafa. “Are you ready for action?”

“As ready as I can be.”

“It shouldn’t be threatening to anyone for us to have our phones out as we’re entering the building,” Pablo said. “We’ll immediately know exactly where baker and Gibson are. I say let’s head straight toward them.”

“We need to remember they’re smart people and we need to plan accordingly.”

Pablo grunted. “We’re smart people too, and we have a lot of experience in situations like this. We’ll assess the situation when we get inside the building.”

Rafa glanced at Pablo. “It won’t hurt anything to have a

plan on what we'll do if the door is locked. It could be a steel door, and we might not be able to shoot the lock off quickly. It might also have some sliding bar locks toward the top of the door. I wouldn't put that past them."

Pablo looked at Rafa. "You make some good points, but I'm not sure what we can do if the door is locked and it's a steel door."

"Maybe Bolivar will have some good ideas," Rafa said. "We should check with him before we go inside. Maybe there's some type of explosive we can use that come's in a miniature package."

Pablo nodded.

"You make the call," Rafa said. "You're better at discussing these types of things."

Both men slowed and moved over to a bench in an area outside the entrance.

Pablo removed his phone from a pocket and punched a digit.

Chapter 47

At Leon's mansion in Germany, Bolivar answered his vibrating phone.

"We're almost to the entrance of Tuxtun," Pablo said. "Rafa and I have something to check with you on before we go in."

"Go ahead."

Pablo explained their concern.

"They shouldn't be expecting you and I doubt they'll have their door closed, and if they do, it's probably just a regular door." Bolivar paused. "The two of you should have no trouble kicking it in."

"Any suggestions on what we do then?" Pablo asked.

Bolivar hesitated for a moment. "You guys are experienced assassins, what would you normally do?"

"We would normally dive for the floor inside the office or pull back beside the wall outside the doorway," Pablo said. "We would, of course, be on opposite sides of the doorway with our backs against the wall and have our pistols in our hands ready for action."

"If you need to kick in the door, how do you decide which

option to use?" Bolivar asked.

"We'll need to act quickly so we'll decide which one we'll use before we kick the door in," Pablo said.

"Sounds like a good enough plan. Why are you wanting any advice from me?" Bolivar asked.

"We don't want to overlook any new techniques," Pablo said. "We just thought it would be a good idea to check with you in case you had some other suggestions."

"Nothing new to suggest," Bolivar said, "however, if you should find the door closed and it's a steel door, then leave. You can get out as easily as you got in. We'll fix you up with the appropriate plastic explosive and you can go back." He paused. "As I mentioned earlier, I don't think the chance of Ralph Gibson having a steel door to his office is very high, but, if he has one, that will be the backup plan."

"Okay," Pablo said. "We're ready to go. We'll keep you posted as is necessary."

Pablo and Rafa left the bench and again walked toward the entrance to the Tuxtun headquarters building.

"We don't just have the advantage here: we have an overwhelming advantage," Rafa said.

Pablo nodded. "The security alarms won't detect us, and we'll know exactly where they are." He looked at Rafa. "That's quite an advantage to have."

Rafa glanced at Pablo laughed. "And we have a good backup plan just in case they have a steel door."

"We're world class assassins, probably the best there is," Pablo said. "We don't need things to be this easy. They could have used some of the other guys who don't have all of our skills."

Both men continued toward the entrance.

Chapter 48

In his plush office as CEO of a large global corporation in Geneva, Switzerland, Zeus sat with Odin and Jupiter at the round conference table in his office.

The door was closed, and the room had been swept for all types of electronic bugs. The sensors in their security system for detecting the bugs were constantly being updated by some of the leading hi-tech experts in the world.

Each man studied the laptop in front of him which had a live picture of Pablo and Rafa approaching the front entrance of Tuxtun.

"Looks like the assassins from Belzex are about to do their thing," Zeus said.

Jupiter nodded. "I'm glad we have the technology to check on most anything we're interested in."

Odin looked up from his laptop and laughed. "Yeah, I'm certain our technology is the best in the world. It's impossible to know that for sure, but I think it's true."

"Any second thoughts about allowing them to kill Gibson and Baker?" Zeus asked.

"It might be just as well to allow them to proceed," Odin

said. "We might have some problems with them later, and there's no loss to us in them being gone. In fact, it might even be a benefit to us. If Belzex gets the extra land they'll substantially increase their revenue from drug trafficking."

Zeus nodded. "Yeah, you're right since we have a way of peeling off some of that profit and putting it in our coffers."

Jupiter leaned forward. "We also have another benefit in allowing this to proceed." He paused as Zeus and Jupiter turned toward him. "We'll get to see how well the technology Belzex has at this time works."

"We know they're trying to catch up with us," Odin said. "Since we know NSA has just updated their detection technology, we'll be able to see if Belzex is making any progress."

Zeus and Jupiter both nodded.

There was a brief silence.

"Hugo thought he had the best technology in the world," Zeus said. "But he was wrong, and he paid the price." He glanced at the other two men. "We know Belzex has the same goal. You're right, it will be interesting to see how this goes."

Odin gave a slight shrug. "It will be a good test of their current capability, but we'll need to continue to keep a close eye on their progress."

After a short pause, Zeus leaned forward and said, "Yeah, if they don't succeed right now, we can be sure that Belzex will continue their work to come up with the ability to make someone completely invisible to any security system."

Odin narrowed his eyes and pressed his lips together. "We know Leon was on the same track as Hugo before and we know he has the same genes as Hitler, just like Hugo did. I think we can be positive Leon is on the same track again."

"That's beginning to bother me a lot," Zeus said. "We should make sure we eliminate Leon along with Belzex this time, and we probably will need to do it pretty soon."

Jupiter nodded. "Like we keep saying: he obviously didn't learn his lesson."

"He could even end up with better capability than Hugo had," Odin said.

Zeus rubbed his chin slowly. "We should definitely plan to go ahead and eliminate him soon."

Odin turned slightly toward the other two. "And remember, some of the same Chinese engineers who worked with Wan Lu in Hugo's group now work in Chen Wan's company. Belzex has the advantage of all that experience."

"And they'll have a lot of additional funding from Bolivar Ortiz's and Jorge Soto's large drug trafficking operations, and, if Belzex gets the extra land from Tuxtun, their revenue from drug trafficking is about to get even larger," Zeus said.

"A huge amount of extra funding comes in handy when you want to accomplish big things quickly." Jupiter narrowed his eyes and leaned toward Zeus and Odin. "They need a lot of power to run their new hi-tech datacenter, and they're going to have it with the new powerplant they're building."

Jupiter rubbed his chin. "Yeah, and if they get the extra land, they'll have the funding to build the best possible powerplant."

There was a brief silence.

"We're lucky Jake Bolton is still working with Leon," Odin said. "He knows what happened to Belzex before and he knows what happened to Hugo. He should have a keen appreciation of how powerful we are."

Zeus nodded. "He's definitely the best person to convince Leon to not mess with us." He paused. "As far as Belzex goes, we'll continue to keep a close eye on them."

Zeus looked at Odin. "And speaking of that, what's the status at Tuxtun?"

Odin looked down at his laptop. "The assassins are ready to go in the entrance at Tuxtun. We'll see how they do. We might soon be saying goodbye to Baker and Gibson."

Chapter 49

At Tuxtun, Ralph continued the meeting in his office and scanned the faces around him.

Ben, Matt, Judy, Todd, and Steve filled five of the other chairs.

"Well, we know Leon Schulz wants to buy some of our land for growing poppies, and I think he's convinced we're not going to sell it to him at any price," Ralph said.

Ben nodded. "Yeah, you've again rejected his offer of ten million dollars per area. He should be convinced." He glanced at Matt and Ralph. "Which means you had better be on your toes constantly."

Ralph looked directly at Ben. "Thanks to you we have the latest updates in our security system."

"Yeah, NSA continues to be very supportive," Ben said. "It's a mutual benefit. We give them an extra test site for the improvements they're constantly making."

"Did they tell you much about the latest updates they made to our systems?" Matt asked.

Ben gave a short laugh. "I don't get into all that. I defer all that to Kyle and he handles it.".

Matt leaned forward. "I know he has joint meetings with Todd, Dan, and Wendell a lot. They seem to coordinate well, and all our security systems get constant updates. Under the circumstances, we need that." He paused. "Todd told me we got an update this morning."

Ben nodded. "Kyle informed me of that, and he told me NSA had concentrated on making sure our systems could detect any intruders who got inside the building. He mentioned that NSA was aware Belzex had been constantly working on giving their assassins invisibility to the detection systems, and NSA is always working to counter that capability."

"So, it's kind of a Ping-Pong situation." Matt glanced at Ralph. "We know Belzex got an assassin in and out of here the night before our meeting with them."

Ralph nodded.

"Yeah, I remember that well," Ben said. "Hopefully these last updates will put you back on top where your security system will detect any intruders who get in."

"We'll stay on high alert," Matt said. "We know Leon needs the land to grow more poppies so he can increase his revenue in the drug business. So, he's going to try to get us out of the way so he can deal with someone else." He paused. "I think he's the kind of person who won't give up, so we definitely have a major problem to deal with, possibly for a long time."

There was a brief silence.

"Does anyone have a strong opinion on how soon we might be encountered by some assassins and where that encounter might take place?" Ralph asked.

"According to the phone conversations NSA intercepted, Belzex thinks they still have their technology working to make their assassins invisible to all the security monitors. So, they might come here at any time." Ben glanced at Ralph and Matt. "So, stay on ready."

"We couldn't be more prepared for an attack than we are right now," Matt said. "We have our protective vests on, our weapons ready for action, and we've just completed more training to maintain our sharpness."

Steve nodded.

There was a brief silence.

"I had another long conversation with Paul about the joint financial office Belzex established here," Ben said. "He thinks we should go ahead and pay them a visit. He told me he'll let me know later what he recommends on the timing."

"I hope he's discussing this with Joe Frank," Ralph said. "It might be best for the CIA to also be heavily involved this time."

"I think he's coordinating with Joe," Ben said. "I'll keep you informed."

After a little more discussion, the meeting ended, and everyone stood.

Ralph looked at Matt. "If you have time you can stay for a few more minutes. I want to coordinate a few things with you to make sure we stay on the same page as we go forward.

Matt nodded and settled back into his chair.

Chapter 50

A few minutes later, Rafa and Pablo walked into the building along with a couple of other people who they thought might be employees.

No audible alarms went off, and no one seemed to pay any attention to them.

Each assassin glanced at his phone as they proceeded down the long hallway. Each phone showed two bright spots down at the end of the hall. They continued walking toward their targets.

The door was wide open in the room down at the end of the hall.

Things were looking even easier.

"Looks like they have a lot of confidence in their security system," Rafa said. "They think they'll get a warning if anyone unexpected gets into the building."

Pablo nodded. "Apparently they didn't get any warning." He glanced at Rafa. "Are you still okay with the plan we formed if the door to Gibson's office was open?"

"It still sounds good to me." Rafa paused for a second. "Which side of the doorway do you want to be on when we rush into the office?"

Pablo returned Rafa's glance and shrugged. "It doesn't make any difference to me."

"I'll stay on the right, and you stay on the left," Rafa said. "No need to switch."

"Sounds good." Pablo narrowed his eyes and glanced at Rafa. "We'll pick up our speed about two seconds before we get to the doorway and rush through the opening. In the first split second, you'll scan the situation on the right side of the room, and I'll scan the left."

Rafa gave a thumbs-up.

*****

Matt and Ralph had heard the beeps a few seconds ago, and they saw the red light on the wall flashing. The volume on the beeps had been purposely kept low to maintain the element of surprise.

"Looks like we have company," Matt said.

Ralph nodded. "Same plan?"

"Yep." Matt drew his Glock 19 and rushed to the corner directly behind where he was sitting.

Ralph also drew his pistol and rushed toward the other corner along that wall.

Both men were hidden from view by anyone in the hallway.

They both pointed their pistols toward the door and waited.

*****

Pablo increased his pace slightly. "This is what I love to do. I'm ready for action, but this one might be too easy to be fun for us. I like a little more of a challenge."

Rafa grinned and kept even with Pablo. "Same here, but we'll enjoy the success no matter how easy it is."

There were still no audible alarms they could hear.

Even though they were moving slightly faster, they still walked at what could be considered a normal pace in case someone came into view ahead of them or turned around a corner behind them. They didn't want to draw any extra attention from anyone until the very last second, and then it didn't matter.

They were now about ten feet from Ralph Gibson's office, and no one was in the doorway ahead of them. Everything around them was still all clear.

"Showtime," Pablo said.

Both men drew their pistols and picked up their pace. They broke into a run about five feet from the doorway into the office.

Chapter 51

Matt and Ralph were ready as two figures rushed at high speed through the open doorway.

Two muffled noises sounded almost simultaneously.

Both assassins fell forward with their momentum carrying them a few more feet. Within two more seconds they sprawled on the floor in Ralph's office.

Matt walked over to the assassin he had shot and checked the pulse. The assassin was dead.

Noting that Ralph had done the same thing and had given a thumbs-up, Matt said, "Well, I guess we handled this threat well enough."

Ralph nodded just as Steve rushed through the doorway.

"Sorry I didn't get here sooner," Steve said. "I was in a remote part of the building."

At that moment, Todd and Judy rushed into the room.

Everyone stared at the two dead assassins for a few seconds, looked around the room, and then looked at Matt and Ralph.

Steve looked back at the two dead assassins on the floor. "Looks like you took care of things."

"It helps to be prepared," Matt said.

Ralph scanned the faces around him. "Yep, nothing like making sure you're ready to handle something like this. We've had plenty of practice for most any threat we could encounter."

"And it's good NSA provided the update for the software in our security system," Todd said. "We apparently leapfrogged over the latest enhancements to the Belzex capability."

"Who's going to notify law enforcement," Steve asked, while scanning the faces of the people around him.

Ralph removed his phone from a pocket and looked at Matt and Steve. "I'll notify Ben. Steve you call Justin. Matt, you call Vince. "Even though most of us just concluded one meeting, everyone will probably want to get back together and compare notes on the relevance of this latest event."

Chapter 52

In Geneva, Switzerland, Zeus stared at his laptop for a few seconds and then looked up at Odin and Jupiter.

They were all sitting at the round conference table in Zeus' office.

"Well, that didn't go so well for Belzex," Zeus said. "They still have their problem of getting the land they need to grow poppies."

Odin nodded. "It's a good thing we have the best advanced technology available to be able to keep up with everything that's happening."

"Belzex obviously is lagging behind in their technology," Jupiter said. "They thought they could keep Tuxtun's security system from detecting their assassins, but this time they were wrong."

Zeus leaned forward. "Yeah, it's a leapfrog process and we're still superior overall. However, Belzex will probably make fast progress in trying to catch back up to NSA and Tuxtun."

"We need to continue to watch everybody closely and make sure none catch up with us." Odin paused. "We don't want to have to deal with the kinds of problems we'll have if anyone draws even with us."

"And we need to make damned sure no one surpasses us," Jupiter said. He looked at Zeus and Odin. "That could end up being a huge calamity in many different ways."

"We'd certainly need to destroy Belzex if that ever happens with them," Zeus said. "This time, we'll make sure we eliminate Leon along with the rest of the reconstructed Belzex."

Odin narrowed his eyes and looked at his two partners. "I hope if things ever get to that point, we'll still have the capability to destroy them. We can't let them improve too quickly and sneak up on us."

"That's all the more reason to keep a close eye on them," Zeus said. "And we'll have to form a detailed plan real soon on how we'll know when the time is right to shut them down."

Jupiter leaned forward. "Since having the best hi-tech capability is constantly leapfrogging among Belzex and NSA, that's the dilemma we face." He paused and looked toward Zeus and Odin. "Belzex is useful to us right now with all the disruption they're causing which makes it easier for us to stay hidden." He shook his head slowly. "But if they cross the line with all the capability they're developing, we'll need to shut them down immediately."

Jupiter nodded. "And it won't be an easy task to make sure we know exactly where that line is they can't be allowed to cross." He looked at his two partners.

They both nodded.

A brief silence ensued.

Zeus narrowed his eyes. "We'll need to immediately do some more thinking on this subject, and we need all our groups to put their minds to this task too. We all need to be prepared to thoroughly discuss it soon." He paused. "I'll probably have another meeting with Vulcan, Neptune, and Apollo within the next couple of weeks. I'm anxious to hear what they have to say about a lot of aspects of this."

"We might need to finalize some specific plans quicker

than that," Odin said. "Especially with Belzex working hard to develop a lot of new capability quickly to use when they move into their new hi-tech datacenter."

Zeus nodded. "Belzex has proven before they're fully capable of making fast progress. With all the resources they've accumulated in their AI company and their two manufacturing companies, I expect them to have all the capabilities they had before and Hugo had. They could have most of the overall capability we have within a few months."

Everyone nodded.

There was a brief silence.

"The Belzex group probably doesn't know yet exactly what happened at Tuxtun. Should we call Leon and give him the news?" Odin asked.

"It would be good to make the point we're still ahead of them on all fronts, especially the technology front," Jupiter said. "I'll have someone give Jake Bolton a call. He can inform them about what happened. He can also remind them again to be careful to stay in their lane."

"Yes, it will be good to remind them again to stay in their lane." Zeus narrowed his eyes. "The Specter Matrix must function properly for us to accomplish our goals. If any group interferes with that, they must be eliminated."

Chapter 53

At his mansion in Germany, Leon turned and looked at Bolivar. "Have you heard anything from your assassins in the last few minutes?"

Bolivar tightened his lips and shook his head.

Leon glanced at his watch. "Based on the last discussion you had with them, they should have called by now."

"No argument about that," Bolivar said. "Maybe their phone signals are being blocked along with the Tuxtun security monitors. Let's give them a little more time. It's not likely they failed. They might have some reason to check some more things out while they're in the building."

"Maybe we should have waited until we had the next app available in that series," Chen said.

"The app that will allow us to receive the signals we're blocking on Tuxtun's security monitors?" Ivan asked.

"Right," Chen said. "I have a team working on it. We should have it ready in a couple of months."

Leon frowned. "That technology wouldn't have helped them succeed."

"At least we would know everything that happened."

Bolivar glanced at his watch and then looked back at Leon. "We know what time they were going in, and they've had plenty of time to finish the job." He paused. "If their phones aren't being blocked by the signals we're sending, we might have a problem."

Jake Bolton removed his vibrating phone from his coat pocket. He checked the caller ID and answered.

Everyone paused and listened to Jake's side of the conversation. They immediately saw Jake's expression change. He frowned, tightened his lips, and continued to listen.

In a few more minutes, Jake disconnected and looked at the five people around him.

"Was that call from the Hidden Empire?" Leon asked.

Jake nodded.

"Well, what did they say?" Leon asked impatiently.

Jake looked at Bolivar. "Your two assassins are dead, and they didn't kill anyone. The mission failed in every way possible. Our technology didn't shut down their detection capability. Ralph Gibson and Matt Baker were both ready for our assassins." He tightened his lips even more and shook his head slowly.

Bolivar moved both arms out to the side vertically, palms up. "How is that possible?"

"The person who called told me that Baker and Gibson got the alarm that our assassins were in the building and were ready for them. They had no trouble at all."

Leon stared at Jake. "The caller talked a long time. What else did he tell you?"

"All the usual stuff we've come to expect." Jake shrugged. "The Hidden Empire is still ahead of us in technology and they're keeping a close eyes on everything we're doing. They were able to get the signals from Tuxtun's security system, and so they saw and heard everything that happened in real time."

No one commented.

Jake paused and seemed to be deciding what to say next.

"I assume the person who called wanted to rub it in that

they were ahead of us in their AI capability, right? Chen said with a smirk.

Jake nodded slowly. “I was coming to that. I was trying to decide how to summarize the key points in his message. He was quite talkative, and he went over a lot of details in multiple ways for a long time.”

“Was he one of the people you’ve talked to before?” Leon asked.

Jake paused for several seconds.

“I don’t think so,” Jake said. “His voice wasn’t familiar, and I think it would have been if I had talked to him before.” He paused again and then shook his head slowly. “No, I don’t think I’ve ever talked to him before.”

Leon motioned with his right arm. “Okay, so he was a different person than you’ve talked to before. He apparently rambled a bit but go ahead and give us some more details about what he told you.”

Jake narrowed his eyes and didn’t comment immediately. He seemed to be thinking.

Everyone waited for Jake to say something.

After a few more seconds, Leon leaned toward Jake and rubbed his chin. “Well?”

Jake hesitated a moment more while he scanned the faces around him. “He emphasized several times that the Hidden Empire was ahead of us in technology and they were constantly watching us carefully.” He paused. “He kept reminding me that they were okay with what we were doing now because it caused disruption, and they liked disruption because it helped them stay more hidden, but there’s a line we can’t cross.”

Leon grunted. “Well did he make it absolutely clear where that line is?”

“Not really,” Jake said. “I was trying to figure that out. I was trying to remember exactly what he told me.”

“Next time put him on speaker phone,” Chen said.

“I would have done that, but he told me not too at the very beginning of the conversation.” Jake paused. “In fact, he emphasized several times for me not to do it."

"But the Hidden Empire is okay with what we're doing so far?" Leon asked.

"That seems to be the case," Jake said, "but, like I mentioned, he wasn't really clear on many specifics."

"But he made it clear the Hidden Empire likes the disruption we've been causing?" Bolivar asked.

"That was very clear," Jake said.

Bolivar nodded. "It keeps Law Enforcement occupied so they don't have a lot of spare time to research things and make a lot of discoveries about what they're doing."

Jake looked at Leon. "He also emphasized a lot that you had Hitler's genes and naturally had a strong impulse to conquer the world. He told me to remind you again that they won't tolerate that. He emphasized several times that you must make sure you don't give in to that impulse."

Leon smirked. "Yeah, I know, I know. He's still referring to that 'stay in your lane' bit." His glance took in both Bolivar and Jorge. "Are you guys in perfect sync on working with Chen to get our hi-tech datacenter built?"

Bolivar and Jorge both nodded.

Leon laughed. "Guess our hi-tech datacenter, which will in time become a fortress, will become easier and easier to defend as we make improvements over time."

No one commented.

"The Hidden Empire will know where we are no matter what we do. Hugo was well hidden in Germany, and they found him. If they decide to get rid of us, they won't nuke us like they did Hugo because we'll again have a lot of important neighbors," Leon said. "We were smarter than Hugo last time and we'll stay smarter this time."

Chapter 54

Later that afternoon at Tuxtun, eight people gathered at the round conference table in Ralph's office. Ralph, Matt, Judy, Todd, Steve, Ben, Justin, and Vince were all there.

The bodies had been removed, the floor had been cleaned, and everything was as close to normal as possible.

Justin scanned the faces around the table. "Even though conditions have returned to being close to normal, wouldn't it be better to wait until tomorrow to have this discussion?"

"The eight of us don't all get together that often and we have a lot of urgent things over a wide variety of subjects to discuss." Ralph paused and slowly looked around the table. "I'll speak for myself. I'm fine and I would rather dive into those things right now."

Matt nodded.

"Well, you're the ones who've endured the most trauma." Ben gave a slight laugh. "I would assume the rest of us can handle a discussion, no matter how many details need to be covered."

Everyone nodded.

Ralph looked at Ben. "Even though Matt and I didn't need them this time, it's a good thing you and Steve both recommended we all wear protective vests and carry our pistols all the time," Ralph said. "I'm sure we'll need them in future situations."

Ben nodded. "It's a good policy to have right now in our overall situation."

Everyone gave a thumbs-up.

Ralph's glance took in both Todd and Judy. "With both of you working with various law enforcement agencies trying to analyze the cashflow activity in the Specter Matrix, I'm guessing the Hidden Empire is going to come after us too, sooner or later."

Ben, Justin, and Vince each nodded.

"The Hidden Empire doesn't seem to have a problem with any of us right now," Ben said. "Leon and Belzex definitely do have a problem with us. And I'm sure they have a lot more assassins to use." He looked at Matt and Ralph. "In some ways it's redundant, but I think it would be good to review all of your defensive measures again, both at home and at work."

Justin and Vince both nodded.

"The way things are shaping up, none of us want to overlook something that would save our life," Ben said. "This would be a good time to review things and try to make sure we haven't overlooked something."

A detailed discussion for about an hour followed.

Chapter 55

The next day in Geneva, Switzerland, Zeus, Odin, and Jupiter sat at the round conference table in Zeus' office.

Zeus narrowed his eyes and looked at Odin and Jupiter. "All the members of Belzex should be well aware our technology is still superior to theirs. And we need to make sure it stays that way."

Both Odin and Jupiter nodded.

"On the subject of our technology being superior, I just had a detailed discussion with some of our technical team," Odin said. "Even though I have advanced engineering and math degrees, I still didn't fully understand what they were explaining."

Zeus and Jupiter didn't comment. They appeared to be waiting for Odin to continue.

Odin leaned forward. "I, of course, was most interested in how well we're protecting the operation of the Specter Matrix since that's essential to our continued existence."

"Did they mention any problems?" Zeus asked.

" No and they had some good news that I wanted to share with you." Odin grinned. "And don't worry, I'm going to keep my explanation as non-technical as I can make it."

"Good," Jupiter said. "My expertise is in finance. I know some basic information about the technology involved but I'm no expert on all the technical details involved in the global cashflow."

Zeus nodded. "Ditto."

Odin leaned forward. "Bottom line, the good news is that we have a new way now of insuring the cashflow we initiate in the Specter Matrix is not interfered with."

"That's interesting," Zeus said. "If you understood the explanation as to what we're doing, I'd like to hear it."

"We're using a lot more advanced quantum processes with our newest computers." Odin turned a little more toward Zeus and Jupiter. "We're using photons now to transmit financial keys and anyone observing a financial state alters it and sets off an alarm." He paused. "Therefore, even if Belzex and NSA are both using their quantum computers to monitor the activity in all the global financial systems, they won't be able to detect any transactions flowing through the Specter Matrix without setting off an alarm."

"And a photon is a particle of light?" Zeus asked.

"Yes," Odin said. "In many cases now when applying the latest quantum technology, photons are used instead of atoms to transmit data."

There was a brief silence.

Odin gave a short laugh. "I know you guys don't have engineering degrees, but if you'll humor me I'll tell you a little more about photons."

Zeus gestured for him to continue.

"A photon is a specific type of boson light," Odin said.

Zeus raised a hand. "Before you go any further, tell us what boson light is."

Jupiter nodded.

Odin took a deep breath and exhaled slowly. "That was explained to me, but now we're getting into the fringes of what I understood."

"Give it a try," Zeus said.

Odin hesitated for a few moments, seeming to gather his thoughts. "I was told it was part of the fundamental physics of light particles. I remember they mentioned it was a specific line of tactical night vision technology." He paused. "I mainly remember their saying that a photon was a specific type of boson light, and I'll continue with what they told me after that about photons."

"Okay, sorry for the interruption," Zeus said. He motioned for Odin to continue.

"A photon is an elementary particle of electromagnetic radiation that includes visible light and radio waves. It is massless and chargeless." Odin paused. "Our experts emphasized that multiple photons could occupy the exact same quantum state simultaneously, and that's one reason our quantum computers can perform as well as they do."

Zeus and Jupiter both nodded and made no comments.

Odin shrugged. "And that's all the details I retained from the conversation about how we can detect an intruder to our activity involving cashflow in the Specter Matrix."

"I think that's pretty damned impressive," Zeus said. "Our technology is likely ahead of everyone else and we need to keep it that way."

Jupiter and Odin both nodded.

"Maybe the knowledge that our technology is superior to theirs will help Belzex stay in their lane and not force us to destroy them," Odin said.

Jupiter shrugged. "That didn't work with Leon last time and it didn't work with Hugo."

"Either Hugo thought that his technology was superior, or he thought he would find a way to destroy us before we could destroy him." Zeus shook his head slowly and leaned forward. "Hugo had Hitler's genes along with a big ego which overshadowed his logical thinking. I'm convinced he thought he could find us and destroy us before we could destroy him." He paused. "We know Leon also has Hitler's genes. I guess the urge someone with those genes has to conquer the world is just too powerful to resist."

There was a brief silence.

"Back to the current technology subject," Zeus said. "We've been in existence for a long time and our forefathers probably never dreamed we would be in wars about AI superiority." He gave a short laugh. "That's only been in the picture for a short time, but it has a lot of momentum, and, as we go forward, we definitely need to contend with it."

Jupiter grunted. "Yeah, being a descendent of Hitler, Leon probably can't resist trying to conquer the world no matter how hard he tries. We know Hugo couldn't resist the urge."

"The genes both men inherited obviously created an irresistible thirst for ultimate power," Odin said. "As of now, I don't think Leon will heed any of the warnings we give him. I think we can plan on him being a real danger to us, especially if he keeps improving his AI capability."

"I think that's correct," Zeus said. "He's beneficial to us right now, but that's going to change. Like we've mentioned before, we'll keep a close eye on what they're doing. When Leon and Belzex cross the line we'll need to destroy them, and that will probably be in the not-too-distant future."

Odin leaned forward. "Unlike Hugo, Leon again chose to put his hi-tech center in a populated area. That creates an entirely different problem for us."

Jupiter nodded. "Yeah, we certainly can't use a nuclear missile, even a smaller tactical one. We'll do something similar to what we did last time with Leon, along with some necessary improvements."

A brief silence ensued.

"Many things could a lot different, as well as much more difficult for us going forward," Odin said. "We have some big challenges to face right now. Belzex is in the process of improving their AI capability rapidly, and they again have accumulated all the right people and resources. We'll need to make a huge effort to stay ahead of them."

Zeus looked at Odin. "You've been overseeing our activity in that area. Do you need some help?"

"I'm okay right now," Odin said. "But I might need some more help soon. I'll keep you posted."

Zues nodded and didn't comment.

"With a new upscale AI company joined with two upscale manufacturing companies and funded by a tremendous amount of illegal drug money, we know they have tremendous capability," Jupiter said. His glance took in both Zeus and Odin. "We've just had some great discussions, but we need to have a lot more soon. We need to jointly figure out the best way to supercharge our efforts to remain ahead of both Belzex and the NSA. Both might be gaining on us fast."

"So do you agree to increasing the funding for our AI resources and giving Odin some more help immediately?" Zeus asked.

"Definitely," Jupiter said.

Zeus narrowed his eyes. "We have a very fine balance we need to keep in mind. Sudden changes with massive funding can attract unwanted attention in a lot of law enforcement areas."

Odin tightened his lips and stared at Jupiter.

"We've been doing similar things for a long time," Jupiter said. "We know how to make the changes to get the benefits we want without losing any of our secrecy."

"Yeah, I know." Zeus leaned forward. "I figured it wouldn't hurt to remind you about that. It's easy to get into too big of a hurry when critical things are at stake. We must always be careful to maintain every bit of our secrecy."

Everyone nodded.

Chapter 56

The next day, at his mansion in Germany, Leon sat in his plush lounge area and scanned the faces around him.

All the Belzex leaders were still together.

"I think we can achieve an AI advantage over both the NSA and the Hidden Empire very soon," Leon said. "We have all the right ingredients in place to make rapid gains in all the critical areas." He turned slightly and looked at Chen.

Chen nodded. "My team is developing new applications rapidly. We think we've recovered the capability to make our assassins invisible to anyone's security system." He paused and scanned he faces around him. "We again can depend on any of our people being able to remain invisible to any security monitors anywhere in the world. And we'll be able to see all the activity ourselves in spite of all the signals being blocked for all our targets at any location."

The discussion continued for an hour.

After a brief pause, Leon asked. "What else do we need to discuss before we end this meeting?"

"Our joint financial office in Jacksonville is about to get a visit from Law Enforcement, "Chen said.

"How soon?" Leon asked.

"They haven't set the exact time and date yet, but it's clear from their phone conversations we've intercepted they're going to do it soon," Chen said.

Leon nodded slowly. "Let me know as soon as you find out when the joint financial office will get a visit. We'll prepare to give whoever comes a big surprise."

"Do you have something in mind?" Ivan asked.

"I'll wait until I find out who's coming and then I'll decide," Leon said. "If Baker and Gibson aren't in the group, I don't want to waste any good techniques to knock them off."

Jorge shrugged. "You could practice whatever technique you might choose to use later to see how well it works."

"I don't want to do that because they would be prepared for it if I tried to use it again. We've underestimated them too many times, and I don't want to take a chance," Leon said.

Everyone hesitated for a moment and then nodded.

Chapter 57

That afternoon in Jacksonville at Tuxtun, Ralph opened a new meeting with his staff in the main conference room. Ben and Kyle also were in attendance.

Ralph leaned forward toward Ben. "Have you talked to Paul about our planned visit to this new joint financial office Belzex has established?"

Ben nodded. "Paul basically agrees with our thinking. He told me he was having a meeting with Joe tomorrow morning to discuss it. He thinks it's best for the CIA also to be involved. They're meeting in Paul's office to avoid discussing it on the phone. We know the Hidden Empire has the capability to keep up with everything we're doing, and we suspect Belzex might now be at that level too."

Ralph looked at Todd. "I know you've put a lot of time into it. How are we doing on our AI?"

"The NSA is working hard to have the same capability as the Hidden Empire; we think we're close. We have all the updates on our system here at Tuxtun and we think we're ahead of Belzex," Todd said. "We don't yet have the funding or the resources they do, but all that is improving." He paused. "Kyle and I have been concentrating mainly on defense and in trying to encrypt our phone conversations better. I think we're making a lot of progress."

Kyle nodded. "We're indeed making a lot of progress. I'm spending a lot of time with the Science and Technology Branch of the FBI. They're spread over four locations, and some people come to our facility in Jacksonville for meetings and sometimes I go to one of their locations."

"Where are those locations?" Judy asked.

"Quantico, Virginia, Clarksburg, West Virginia, Fredrick, Maryland, and Huntsville, Alabama," Kyle said. "They have different specialties at each location."

Kyle took a few minutes to provide a lot of details.

"I've been working closely with him on the advanced biometric technology which is a specialty at the Clarksburg, Virginia location," Todd said.

Matt's glance included Todd and Kyle. "Could you give us some details?"

Todd gestured to Kyle.

"The FBI's "advanced biometric technology" is mainly based on the Next Generation Identification (NGI) system, which serves as the world's largest electronic repository of biometric information," Kyle said. "The NGI system replaced the IAFIS system and expands law enforcement capabilities beyond traditional fingerprints to include facial recognition, iris scans, and palm prints."

"What's the IAFIS system?" Ralph asked.

Kyle gave a short laugh. "Oops, sorry, guess we can't expect everyone to be familiar to all the acronyms." He paused. "It stands for Integrated Automated Fingerprint Identification System. It's a massive, computerized national fingerprint and criminal history database maintained by the FBI's CJIS Division. It allows law enforcement agencies to quickly search, store, and share digital fingerprints and biometric data to identify suspects and conduct background checks."

Todd looked at Ralph. "The CJIS Division is the Criminal

Justice Information Services Division. I figured you might be asking about that one also."

Ralph smiled and gave something like a salute. "Thanks for thinking of me. I haven't gotten into any discussions involving any of those entities, so I wasn't familiar with the abbreviations."

Todd nodded.

Ralph scanned the faces around him. "Any more unfamiliar abbreviations you can think of that might come up in any more related discussions?"

"BTC might be one," Todd said. "It stands for Biometric Technology Center. It's on the campus there in Clarksburg."

Ralph gave another one of his partial salutes. "Let's go ahead and talk some more about some things we need to be familiar with."

The discussion went on for another thirty minutes.

Ralph looked slowly around the room. "Any other comments or questions?"

Matt glanced at Kyle. "Do you know if any of the locations you mentioned have quantum computers?"

"I don't know," Kyle said. He looked at Todd.

"Same here." Todd shrugged. "We'll check on that and let you know."

Matt nodded and didn't comment.

Steve looked at Kyle. "I'm assuming the CIA is doing a lot of the things all of you in the FBI are doing."

"Yep, in their own way, they're involved in everything we've been discussing," Kyle said. "They have the Directorate of Science and Technology (DS&T) and are studying all the subject matter we're studying."

Kyle took a few minutes to give more details on many of the technology subjects the FBI was jointly studying with the CIA.

There was a brief silence.

Ralph turned toward Judy. "I know all our business ventures are doing well, and that's a good thing so we can keep concentrating on what we're doing with Law Enforcement." He paused. "Any new bright spots you would like to share with us while we're together.?"

"Our legal drug business is really taking off. We just had a record crop of poppies, and our sales of Morphine and Codeine products have more than doubled in the last three months," Judy said.

Ben narrowed his eyes and glanced at Matt and Ralph. "Leon is really going to want your land now. You need to stay prepared for a lot of new threats."

Matt looked at Ralph. "We can spend more money on AI, and we should start doing that immediately. We can also hire some more people and increase our research."

Ralph nodded. "In addition to competing better in the business environment, we need to do that as a defensive measure to better protect ourselves."

Everyone gave a thumbs-up.

Chapter 58

The next morning in the Washington DC area just off Chain Bridge Road, Joe Frank the CIA Director met with Paul Becker, his counterpart at the FBI.

"Thanks for coming over," Joe said. He rose from his desk, shook hands, and waved Paul toward the round conference table in a corner of his office. "I thought it was best to discuss this bit about the new joint financial office in Jacksonville face-to-face. No need to take any chances on our phone call being intercepted. We don't want to have that huge disadvantage."

"Yeah, the Hidden Empire has the technology to listen to everything we say if we talk on the phone, and I suspect Belzex does too."

Paul slowly pulled out a chair from under the table and slid into it.

Joe followed the exact routine.

"The Hidden Empire might even be able to pick up our conversation here at the table," Paul said. "I don't think Belzex is quite that advanced yet."

Joe shrugged. "I don't think the Hidden Empire cares if we eliminate the joint office. But anyway, it's best to be as careful as we can on these types of matters. And we're doing all we can within reason."

"Have you swept your office for bugs?" Paul asked.

"Do it at least once a day. Since you were coming over, I just finished doing it again."

Paul gave a short nod and then asked, "What's your thinking about some group paying a visit to this new joint office Belzex has set up in Jacksonville?"

"I'm okay with it," Joe said. "I'm guessing that's what you recommend."

"Have you done any thinking about whether or not any of your agents should participate in it?"

Joe rubbed his chin and looked directly at Paul. "I think the CIA should participate. It's the logical thing to do. We all have a stake in this."

There was a brief silence.

"Which of your agents do you suggest should be in the group?" Paul asked.

"Kevin Brown and Butch Reilly. You'll probably need some extra protection."

Paul grunted. "That's quite a talented duo."

"We want the right people in the group when it comes to making sure we don't have a well-organized terrorist organization operating in our own backyard."

Paul leaned forward. " I'll send Dan Maxwell to help with the inspection. He's had experience working with the agents you're sending. I'm sure Ben Fulton, Kyle Rankin, along with a group from Tuxtun will also be involved."

"And I think they've all had experience working with my agents, right?"

"Yep," Paul said.

Paul gave a short laugh. "I think we're going to have a well-qualified group in every way."

There was a brief silence.

"The group should have some good protection," My agents tell me that Kevin and Butch are two of the best warriors on the planet," Paul said.

Joe nodded. "They have that reputation and I know it's well earned."

"Matt Baker and Ralph Gibson aren't too shabby themselves and they've been well tested.," Paul said. "We shouldn't need to worry too much about their safety."

Joe leaned forward. "And it's a good additional factor that they're all used to working together."

"Remember, the security group that was there with Hugo in that joint office he established used a bazooka to try to eliminate our inspection team." Paul shook his head slowly. "But, with our team's experience, I'm sure they'll be prepared for anything."

After the exchange of a few more comments, Paul stood. "Okay, we'll check a few more things and I'll let Ben know to get it scheduled. I'll keep you informed."

Both men gave a thumbs-up.

Chapter 59

The next morning in his mansion in Germany, Leon relaxed in a lounge chair in his den and glanced at Jake.

Jake was sitting nearby.

"I just talked to Chen. They're making rapid progress on the hi-tech datacenter," Leon said. "Looks like it's going to be finished much sooner than expected and it's going to look more like a fortress again. I'm just going to start calling it a fortress."

Jake shrugged. "That's what we called it before and that's what Hugo called his." He looked back at Leon. "That's not surprising that the fortress will be finished early. Ivan and Jorge both have ultramodern manufacturing companies with the best facilities and equipment on the planet. They also have a lot of prefabricated stuff that can be installed quickly. And they're already manufacturing all the specialty items from Chen. It will indeed be a fortress in every way."

Leon nodded and didn't comment.

"Did Chen give you any target date as to when we can move in?"

"Within the next week."

"Wow, that is fast."

"Chen mentioned he Ivan, Jorge, and Bolivar were in complete sync." Leon leaned toward Jake. "They've already started installing a lot of the AI equipment."

"Guess you're anxious to get started using all that constantly improving capability?"

Leon nodded. "My top priority right now is eliminating Baker and Gibson, and that will remain my top priority until I finish the job."

"What about conquering the world?"

"Of course that will continue to be my long-run goal."

"What about right now?"

"Right now, a close second on the priority list will be to discover a lot more details about the Hidden Empire. When the fortress is complete, the additional AI capability will help greatly with all my priorities. "

Jake narrowed his eyes, looked at Leon, and didn't immediately comment.

"Is something bothering you?" Leon asked.

"Are you going to again consider trying to eliminate them?"

"The Hidden Empire?"

Jake nodded.

Leon narrowed his eyes. "If I want to conquer the world, I'll need to do that at some point in time."

Jake shook his head slowly. "I don't recommend you pursue that line of thought for a lot of reasons. For one, you know what happened to Belzex last time and you know what happened to Hugo."

"If Chen is able to increase our power the way I think he can, I don't think the Hidden Empire will be as big of a problem for us as you think," Leon said. "I'm going to be a lot more powerful very soon, a lot more powerful than I was before and a lot more powerful than Hugo was."

There was a brief silence.

Jake continued shaking his head.

Leon stared at Jake for a few moments. "This seems to really bother you." He paused. "Don't you realize how powerful we're going to be?"

"Hugo thought he was powerful too, and he was powerful in many ways. He was well hidden in a secret location, and he had nuclear missiles. You're out in the open and you don't yet have any powerful weapons."

Leon shrugged. "Being in a heavily populated area protects us to a great extent. They won't nuke us."

"It didn't fully protect you last time. You know what happened."

"I'll be smarter this time."

Jake shook his head some more. "They'll find a lot of other ways to get you."

"Ivan will help us get some nuclear weapons from the Russians and we'll be constantly building better offensive power and better defenses in every way. I expect Chen to quickly raise our AI capability to an unbelievable level."

There was a brief silence.

"I still think it's a huge mistake to underestimate the capability of the Hidden Empire," Jake said. "They've been around for centuries, and they obviously know what they're doing in every way. They've honed all their methods, skills, techniques, and capabilities of many kinds to the highest possible level."

"Yes they've existed for centuries, but we're now in modern times with modern capabilities." Leon paused. "Like I keep telling you, we're on the verge of getting a lot more powerful. Just watch for a while and then we'll have this conversation again in a few more weeks."

Jake nodded slowly and didn't comment.

Chapter 60

The next morning in Jacksonville, Ben and Kyle each drove an FBI company car into a visitor parking area in front of a four-story modern office building in a business park a few miles southwest of downtown Jacksonville.

Eight men occupied the two cars.

Kyle and Dan got out of one of the cars and led the way. Kevin and Butch followed a few feet behind.

Ben, Ralph, Matt, and Todd got out of the other car and moved in front of the other group as they all walked toward the front entrance.

Ben walked up to the double glass doors and rang the doorbell. No one came. He turned the doorknob. As expected, the double doors were locked.

Ralph glanced at his watch. "I suggest we don't wait more than ten minutes."

"And then what, break in?" Ben asked.

"I suggest we go back to the office and make a different plan," Ralph said. "We can reschedule and have a SWAT team ready if we need them." He looked at Ben.

Ben nodded.

Everyone checked their watches and waited.

In less than five minutes, one of the doors opened. A tall slim man stood in the doorway and looked at them without expression.

Ben stared at the man for a few seconds and said, "We have a scheduled meeting." He held out his badge.

"I know," the man said. He opened the door wider and motioned for everyone to come in.

Two large heavyset men with pistols on their hips stood behind the tall slim man. They immediately took a few steps back to get out of the way.

After Ben and his group moved inside and stopped a few feet away from the tall slim man, Ben said, "We need to do a routine inspection of your building for general security reasons."

"I'm the project director," the tall slim man said with a slight foreign accent.

Ben introduced everyone.

"Why is the CIA involved in this?" the project director asked.

"You have foreign ownership, and that makes it necessary for them to be involved," Ben said.

The project director nodded slowly and had a questioning expression as he scanned the group in front of him.

"I brought along some experts on various subject matter," Ben said. "They might have some questions for you. You're welcome to conduct business in our country, but we need to make sure you're not a threat to our nation's security."

Matt noted the security guards narrowed their eyes and shifted their positions slightly. Matt shifted his position slightly too, which made it easier to draw his pistol. He also noted Kevin, Butch, and Ralph had also shifted their positions.

After reflecting a moment, Matt remembered that Butch used a SIG Sauer P226, Kevin carried a Heckler & Koch USP, and each of the other six men carried a Glock 19.

Each pistol had a 15-round magazine of 9x19mm Parabellum cartridges. Matt knew, if they needed it, they had plenty of firepower.

The project director glanced at the guards. "It's okay." He motioned for the group of inspectors to follow him.

The security guards seemed concerned, but they moved to the back of the inspection group and followed about ten feet behind them.

The project director narrowed his eyes, looked at Ben, and asked, "Have you inspected any other buildings in this area?" His slight foreign accent was still noticeable.

"Yes and this is a standard procedure by the FBI," Ben said. "And, as I mentioned, since you're foreign owned, the CIA is involved this time."

The project director nodded and rubbed his chin. "We're going to be processing a lot of sensitive information here. We don't plan to allow any public access, not even from Law Enforcement."

"This is a necessary visit," Ben said. "If you don't allow us an opportunity to inspect everything we'll get a search warrant, and you'll be flooded with a large number of armed agents inspecting everything in detail."

The project director tightened his lips and frowned. He narrowed his eyes again , but he didn't comment as he waved everyone forward.

With the security guards following, they walked through a small lobby and into a long hallway.

The project director ushered them into a large room with several long rows of desks.

No people were in the room. Matt guessed they didn't want them to know about all the different nationalities who might be involved.

A large screen and a keyboard were on each desk.

Matt didn't see any obvious computer connections, but he was sure the connections were there. He figured all the screens and keyboards were connected to a large central computer, probably a supercomputer of some type, maybe even a quantum computer. He figured Kyle, Todd, and Dan were studying the area carefully. He would ask them about it later.

Ben looked directly at the project director. "Do you normally have a lot of people in here?"

The project director shrugged. "We have a large group of people who come in on occasion and help us with a lot of technical things that pertain to their expertise."

"Why do you have so many computer stations?" Dan asked. "And what types of things are these people helping you with?"

"They help us with International banking issues during peak hours," The project director said. "Remember we operate worldwide. We want all the necessary transactions to go through on a timely basis. One reason for this office is to provide backup for our needs in some of our global financial centers that operate our systems."

"Global financial systems supported by the World Bank, the WTO, and the IMF?" Kyle asked.

The project director nodded. "Of course, any business handling financial transactions from around the world will need to interact with the entities you mentioned."

"Why don't you have a lot of people working in here right now?" Dan asked.

"That will probably happen soon," The project director said. "We're not fully set up yet. We still have a lot to do at this site to get ready to handle our worldwide business."

"When do you think you'll be fully set up?" Ben asked.

Both security guards frowned and glanced at the project director.

Matt glanced at Ralph, Kevin, and Butch. He figured they could handle any trouble if it occurred. Each of them could have their pistol in firing position in less than a second if necessary. Also, the other four men in the group were well trained to handle situations like this when necessary.

The project director shrugged. "We still have a lot to do and there are some complications. Since we're an international entity, we must be approved by several federal agencies to do business in the United States." He looked directly at Ben. "That approval is still in process."

Ben nodded and didn't comment.

"Also, the Federal Reserve often plays a part in some of our transactions," The project director said. "That's another big reason we need an office here in the United States. We must coordinate closely with them."

Kyle turned slightly and gestured toward the equipment on the desks. "Is all this connected to a large computer?"

"Of course, we can't do our job if we're not online to a superfast computer that's connected to a superfast network." The project director grunted and then walked back toward the doorway that led into the hall.

The security guards stayed about five yards behind the group and continued to follow.

Matt tilted his head slightly and kept the guards in sight. He saw one of the security guards remove a cell phone from a pocket and punch a button on it. The guard then returned the cell phone to the pocket.

Kevin moved up close to Butch and whispered, "We'd better be ready for something."

"You keep a close eye on the one to our left and I'll keep a close eye on the one to our right," Butch said.

Kelvin gave slight nod.

With the security guards now flanking the group, the project director walked down the hallway and gestured toward a lot of individual offices and several large conference rooms as they passed by.

Matt noted that no one was in any of the offices or conference rooms.

"We'll use these later when we get to full operation," The project director said. He flashed a big smile.

No one commented.

The project director turned and started walking back up the hall toward the lobby. "You've seen all our facility, and I've explained why we wanted an office building here. Is there anything else you need?"

"I think we need a little more explanation about a couple of things," Ben said. He glanced at the rest of his group.

The other seven men in his group nodded.

"Such as?" the project director asked suspiciously. He turned slightly toward Ben.

Ben gestured to Kyle.

"Why do you have devices in every room to intercept various types of communication signals, Wi-Fi, Bluetooth…." Kyle asked.

The project director stepped closer to Kyle. His eyes quickly narrowed as he asked, "What are you talking about?"

Kyle gestured toward Dan and Todd.

"The three of us are well trained in these types of things." Kyle turned fully toward the project director. "Although you have them well hidden, we were able to detect them."

Todd nodded. "We're experts on this technology and it works well for fifteen miles. You can choose the direction you want to scan and right now it's aimed toward the FBI Headquarters on Gate Parkway."

"How can you tell?" the project director asked.

"There is a small prong on each device that you can turn to point in any direction," Todd said. "Right now, all the prongs are turned toward the FBI Headquarters, which is only ten miles away."

"I know nothing about that," the project director said. He made a quick gesture to the guards as he turned and ran toward a nearby exit door as fast as his thin legs would carry him.

Each guard moved his right hand toward the holster under his left arm. Three more people came around a corner at the back of the hall and charged toward them, pistols in hand.

Butch and Kevin shot the two guards close to them before either guard could get his pistol out of the holster .

Matt and Ralph shot two of the three people charging toward them before they could get their pistols into firing position

The other man charging up the hall started firing.

Matt heard some bullets zing by and strike the wall at the entrance.

That was the only chance the charging assassin got. He quickly fell to the floor under a hail of bullets from the inspection group.

All eight men in the inspection group had their pistols pointed toward the downed assassin and as soon as the din of noise from the bangs subsided, Ben punched a digit on his phone.

"I'll get the right people out here to secure the building and do the cleanup," Ben said. "I'll talk briefly with the leader and then we'll leave."

"It's lucky for us that guy wasn't a good shot," Butch said as everyone made a quick search of the area.

Matt nodded. "And it's a damned good thing we were good shots."

Ben waved everyone to the exit the project director had used. After grabbing the handle, he pushed open the door and went outside.

The group quickly spread out and scanned the area in all directions. No one was in sight.

"I figure the escape had been well planned and we should waste no time looking around anymore," Ben said. "Too bad we were all preoccupied with putting down the assault when he made his escape."

Everyone nodded.

"Okay, we've done our job," Ben said, as some approaching sirens pierced the air. "As soon as I have a short conversation with the arriving agents, we're headed back to the office."

Chapter 61

The next day in Germany, Leon sat in a lounge chair in the spacious den in his mansion and reached quickly for his ringing phone.

He answered and listened for a few seconds.

"Do you think we can ever reopen that office on that site?" Leon asked.

"No, everything is lost," the caller said. "The building and all of its contents have been impounded."

Leon listened for another minute and disconnected. He paused for a moment and then looked at Jake who sat in another lounge chair close by.

"I'm guessing that was the project director of our joint financial office," Jake said.

Leon nodded. "Yep, I had one of our fastest jets fly him over here. He gave me all the details of his experience. I don't think there's any need to ever try that again. We'll have to do without a lot of face-to-face meetings with the Federal Reserve."

Jake nodded slowly and waited for Leon to continue.

"Right now, we'll concentrate on getting everything ready for us to move into our hi-tech fortress as soon as possible. That's where we need to be operating from."

Jake leaned toward Leon. "What is the project director going to do now?"

"I asked him to help expedite the completion of our hi-tech fortress. He has some valuable skills we can use and he's now fully available."

"What are those skills?"

"He has a lot of financial skills along with excellent electrical and AI skills. That's why he was made the project director of the joint financial office."

Leon started to say more when his phone rang. He noted it was Ivan, and he answered. He selected the speakerphone option for the conversation.

Jake paused and listened.

"Things are going well as we continue getting our hi-tech fortress built. I just thought I would give you a status report," Ivan said. "I figured you might like some good news for a change."

Leon laughed. "You're right about that, keep going. I'll take all the good news I can get."

Ivan spent the next twenty minutes explaining the capability they were going to have once they all occupied their new facility.

"And how is the project director from the joint financial office working out?" Leon asked.

"He's a big help," Ivan said. "That's one reason I decided to call. He's currently supervising the installation of a large section of a wall on the first floor. We can call it a hi-tech wall because it has all the advanced technology in it to connect with a wide range of our devices without us having to go to a lot of trouble later to get everything connected and working properly. However, everything needs to be placed in the exact right position for all the equipment to function properly. Your former project director for the joint financial office is now taking care of all that."

"I didn't know he had that particular skill."

"He's quite an expert, and he's familiar with the manufacturing business as it pertains to construction of prefab buildings."

"You're constructing our fortress as a prefab building?" Leon asked with some emotion obvious in his voice.

"Don't worry. Jorge and I discussed all this thoroughly. We both have experts on constructing prefab buildings. The material we use is stronger than a regular wall and we can construct the building much quicker."

"I hope you and Jorge are correct about the material you're using being much stronger."

"We are. You can relax. We know what we're doing and each of our companies has done a lot of testing."

"I assume all of your people doing the testing got together and compared notes?"

"Absolutely," Ivan said. "They discussed everything for over twelve hours during their last meeting. They carefully analyzed all the data, rechecked all the facts, and ended up agreeing on everything. That's unusual, but it demonstrates how good our process was for testing and analyzing all the elements involved, especially the pre-installed hi-tech elements in the walls."

"Is Chen there with you?"

"You better believe it," Ivan said. "He's personally interacting with Jorge's and my construction experts as well as his technology experts. Jointly, they have everything planned out just the way you're wanting it. I think we're going to move into the fortress even sooner than we mentioned last time we talked about it."

"It's a good thing you got the foundation finished so quickly last week. That was a great start."

Ivan laughed. "I keep telling you Jorge and I have advanced

methods. You need to start believing it."

Leon was quiet for a few moments and then he said, "You need to remember that I'm in charge of Belzex. I need to be skeptical and make sure everything is done correctly."

"You're going to be very pleased."

There was a brief silence.

"Okay, so when exactly does it look like we'll be ready to move into the building?"

"No later than next week, maybe sooner."

"That's not very exact."

"A couple of variables need to be worked out, but trust me, it's going to be very soon."

Leon hesitated a moment and then said, "Keep up the good work and keep me posted."

"Will do."

A brief silence ensued.

"I know you're busy, but I have another subject to address while you're on the phone," Leon said.

"What's the subject?"

"Nuclear weapons."

"What do you want to know?"

"Do you still have all your contacts in the Kremlin?"

"Yes, and I have to be very careful since the KGB supports the Hidden Empire."

"Do you know why they do that?"

"The Hidden Empire gives them trillions of dollars every year."

Leon grunted and didn't comment.

"Do you have a specific question?"

"Can you arrange for us to get some small nuclear tactical missiles at our hi-tech fortress?"

"Since the Hidden Empire won't allow you to conquer the world. Why do you need them?"

"Extra insurance if some country gives us a hard time on something."

"Do you think the Hidden Empire will allow you to nuke

another country?"

"In some cases, they might. Nuking another country for any reason will certainly create a huge disruption in the world, and I don't think the Hidden Empire will mind that disruption at all."

Ivan didn't comment.

"Well, can you do it?"

"After we get established in our hi-tech fortress, I'll check into it. If the possibility of getting one looks good, I'll give it my best effort."

"Another thing, I would like to get a small tactical explosive device shipped to Jacksonville, Florida. It doesn't need to be a nuke, but it needs to be powerful enough to destroy a city block when it explodes."

"I can guess what's that's for."

"Yep, if we can't do it any other way, we'll use it to kill Baker and Gibson. We'll make sure they're at work and then we'll wipe Tuxtun Industries off the map."

"Don't you think that's overkill?"

"I said if we can't do it any other way."

"Okay, I think I can work that out for sure. Expect a suitable explosive, maybe even a suitcase-nuke to arrive in JAXPORT in a few days," Ivan said. "Make sure you're prepared in every way to receive it."

"Wait until we're established in our hi-tech fortress with every aspect of our AI working. That way I can track every bit of the activity, and I can do some things to prevent the explosive from being detected."

"You got it."

After Leon disconnected, Jake said, "You tried that before and it didn't work."

"This time will be different," Leon said. "We'll have a lot more powerful AI, and we'll control the environment around the nuke better."

Chapter 62

The next morning in Jacksonville, Ralph, Matt, and the rest of the group from Tuxtun sat with Ben, Kyle, Wendell, Dan, Vince, Justin, Paul, and Gary around the large table in the hi-tech conference room at the FBI Headquarters.

Matt was glad Paul and Gary were there. Having the FBI Director and the head of the Special Investigations Unit for Domestic Threats present would make for a more productive meeting.

"Well, I think we've eliminated the joint financial office Belzex established and we don't think they'll try to resurrect it," Paul said.

Ben nodded as he glanced at the group from Washington DC. "I'm glad all of you could fly down here for the meeting. We have a lot of sensitive things to talk about and it's always better to do it face-to-face, and we don't want to have any of our conversation intercepted."

"The Hidden Empire will probably still be able to do it," Gary said. "They have the technology to somehow eavesdrop on personal conversations, not just intercept phone calls. And we know they can easily get live pictures of a lot of various activities."

Wendell removed his vibrating phone from a coat pocket and pressed a button. After listening for a moment, he disconnected.

"And on the subject at hand, NSA just intercepted an important phone call," Wendell said. "Russia is planning to ship a suitcase nuke into JAXPORT."

"How soon?" Ben asked.

"It's not clear yet exactly when it will arrive," Wendell said. He looked at Matt and Ralph. "Your names were mentioned in some of the conversations that were intercepted. Joe has all the necessary people on high alert. They'll monitor the situation and keep everyone posted."

Ralph looked at Matt. "If that nuke is meant for us, they're again taking extreme measures to eliminate us."

"It didn't work last time," Matt said.

Ralph shook his head slowly. "I'm sure it's Belzex and I guess they're going to keep trying."

Ben pressed a few buttons and then removed a laser pointer from a pocket and highlighted a screen on the wall displaying a map of JAXPORT. He then used the laser pointer to highlight some equipment on one of the docks.

"Remember we installed sensors that use our latest technology to detect many things, including nuclear material and illegal drugs," Ben said, "I don't have the knowledge to explain all the technical details, but it's proven to be very accurate in the extensive testing that was conducted."

Gary nodded. "We have all the latest updates from NSA and they're ready to install on all the detection devices." He looked at Ben. "Do you still have the handheld devices for detecting drugs and nuclear radiation?"

"I still have three of them," Ben said. "We'll make use of them again."

Gary leaned forward. "Wendell will get them updated for you."

Wendell nodded. "They have a range of over a hundred yards. I'll get them updated as soon as this meeting is over. We have a lot of flexibility getting them into correct positions." He paused. "The software will be updated, and they'll be very useful for this situation."

Paul looked at Ben. "Just like last time, you're in charge of the task team. I'll let you take it from here."

Ben nodded as he took a deep breath and exhaled slowly before he started talking.

"I want to use the same team as we did before. At least we've had experience with this exact situation." Ben looked at Matt and Ralph. "Are you still game?"

After he saw a couple of nods, Ben used his laser pointer to highlight a spot on the map. "Here's where I recommend we set up at the port when we try to intercept the nuke." He glanced at Matt and Ralph. "With your combined knowledge and skills, you'll be the best ones to go with me. Hopefully our devices will have better detection capability this time."

"Since this directly affects Tuxtun we're glad to help," Ralph said.

Matt nodded.

"Like last time, I'll have a fully equipped SWAT team standing by," Ben said. "We're not going to take any chances with this situation."

Wendell looked at Ben. "With the new Software, hopefully you can detect the nuke this time and the SWAT team can take care of making sure it's transported properly to the right location."

"They've all had the proper training," Ben said.

Paul turned slightly toward Ben. "We have a lot of things going on, but we'll keep a strong focus on your team intercepting the nuke as soon as it gets to JAXPORT."

"We have the right skills and equipment and we're ready for action," Ben said.

Chapter 63

Leon sat at the round conference table in his office at his mansion in Germany and looked at Jake Bolton who sat across the table from him.

"So, you had another contact from the Hidden Empire?" Leon asked.

Jake nodded. "They want to talk directly to you. The messenger told me you can use the same instructions you've used before."

After Leon retrieved the instructions he had filed, he activated his phone and called. He immediately heard a male voice at the other end of the connection.

"Is this Leon Schulz?"

"Yes."

A few seconds of silence ensued.

"Okay, we've validated your voice pattern," the man said slowly.

Leon waited.

"It looks like you didn't wise up and you're again planning

to gain control of the world," the man said. "We've made it clear that's not allowed. We've warned you repeatedly about that."

"I only want to make a lot of money and defend Belzex when necessary. As we go forward, some countries might try to attack us," Leon said.

"Remember, we control the events in the world we consider important, and we have the technology and expertise to do that. We're very well organized and we're very powerful," the man said. "We're watching every move you make. We'll allow you to go forward with your plans only if you stay in your lane."

"Do you think I'm getting out of my lane?"

"You're on the edge, the very edge."

"Is there something specific bothering you?"

"You're getting a nuke. You're trying to accumulate too much power."

"I told you why I'm getting it."

"You're not telling the truth. We know you want to use it to eliminate Baker and Gibson."

There was a brief silence.

"If you try to take over the world, we don't think you'll succeed," the man said.

"I'm not trying to take over the world, and I thought you liked the disruption I'm causing."

"You're right that we like the disruption, but we don't like you using the nuke in the United States. There's a line you're not allowed to cross."

"Using the nuke is crossing that line?"

"Yes."

There was another brief silence.

"Do you have any objection to my trying to eliminate Matt Baker and Ralph Gibson some other way?"

"Not necessarily."

"What do you mean?"

"By trying to eliminate them you'll create a lot of

disruption, and we like that. The disruption helps us keep everything we're doing as secretive as we need to keep it."

"I'm aware of what you just mentioned."

"So, try to eliminate Baker and Gibson some other way and things will be okay."

"Do you have any suggestions?"

The man laughed. "I think you have plenty of methods you can use."

"You're telling me I can figure it out myself?"

"You could say that."

"Why can't you suggest something?"

"I have my reasons."

There was a brief silence.

"So, we can go ahead with our work as long as we lay off trying to use the nuke?" Leon asked.

"Yes," the man said. "We'll keep an eye on what you're doing and get back to you if you cross the line."

"And the line again is?"

"Using the nuke and trying to be powerful enough to eliminate us, of course."

"So, I can keep doing what I want to do unless I hear back from your organization?"

"That's correct," the man said. "If I need to have another discussion with you, we'll use a different procedure. Jake will receive an envelope with instructions. There will of course be a different phone number in the envelope and some different instructions."

There was a brief silence.

"Okay," Leon said. "I get the picture. And for now, I'm going to cancel the nuke and continue forward with my work."

"Remember, don't get out of your lane."

Leon heard the line go dead. He then looked at Jake. "You heard everything okay?"

Jake nodded.

"What do you think?"

"I think you should call Ivan right now and take any actions you need to completely shut down the plan for using the nuke." Jake looked directly at Leon. "And I mean completely shut it down."

Leon grinned. "Maybe I'll go ahead and get the nuke and store it some place."

"The Hidden Empire will know you did that and they won't like it."

"So what." Leon shrugged. "We need to keep preparing to meet the big challenge of taking over the world. We'll move into our hi-tech fortress real soon and then we'll really get busy."

"I think the Hidden Empire gave you a clear warning not to do that."

Leon shrugged again and made no comment.

"What if the Hidden Empire gets back to us real soon and tells us we've gotten out of our lane?" Jake asked.

"We'll deal with that when it happens," Leon said. He flashed a bigger grin.

Jake shook his head slowly and stood to leave.

Chapter 64

The next day at a secret location in Vienna, Austria, Zeus sat with Odin and Jupiter at a round conference table in a secluded room.

The door was closed, and the room had been swept for bugs.

"Having our meetings at different locations around the globe seems to work well for us," Zeus said. "It's a little bit of trouble but I think it's worth it."

Odin looked directly at Zeus. "No argument from me."

Jupiter nodded.

"Now that we've given Leon another warning, we must continue to keep a close eye on what's going on with Belzex," Zeus said.

"Belzex is moving forward quickly with their global plans." Odin leaned forward. "Leon obviously is still going to try to conquer the world and that means trying to conquer us too."

Odin nodded. "It's a good thing we have the best advanced technology available to keep up with everything happening."

"Yeah, we're still superior," Zeus said, "but Belzex is making fast progress. I don't think it will be much longer before they get out of their lane." He paused. "With all the resources in their new AI company and their two manufacturing companies, I expect them to be close to having our capability within a month."

"Do we have anyone constantly monitoring their conversations?" Jupiter asked.

"Yes," Odin said. "I have a team doing that. They keep me well informed, and I'll keep both of you informed on anything I think is significant."

A brief silence ensued.

"I'm more worried about Leon than I was Hugo," Zeus said. "The way things are going, I think Leon is going to end up with more capability than Hugo had."

Odin nodded. "We know the urge to conquer the world is in his genes, and I'm now convinced it's irresistible."

"So where does that leave us?" Odin asked.

"We like his creating all the disruption," Zeus said. "Which is why we've been rather lenient." He narrowed his eyes. "But I think we need to go ahead and eliminate him and his whole operation before it's too late."

"How soon?" Odin asked.

Zeus grunted. "We don't want to make any mistakes this time. We won't be in too big of a hurry. We'll carefully plan exactly what action to take and when to take it."

Chapter 65

The next morning in Germany, Leon gathered with the other leaders of Belzex in a lounge area of his expansive mansion.

"We're way ahead of schedule," Ivan said. "All the finishing touches aren't completed yet, but we can move into the hi-tech fortress tomorrow."

"Will all the advanced features of our AI be ready to use?" Leon asked.

"Yes," Ivan said. He glanced at Chen.

Chen nodded.

"And we've even completed the bomb shelter that you insisted on having," Ivan said. "We've ensured it's identical to the one you had below your previous fortress."

Leon gave a short laugh. "Yeah it worked well and saved my life." He noticed a few curious expressions and paused.

"You used it?" Jorge asked.

"I guess I should explain that to you," Leon said. "One day I was trying to remember if had stored some particular items in the bomb shelter." He paused. "I couldn't remember so I decided to go in and check. That's when the Hidden Empire set off the explosives that destroyed the hi-tech fortress."

"Did you have any injuries?" Ivan asked.

"Not while I was down there," Leon said. "I sustained a few minor injuries getting through all the rubble to get out of the area."

Bolivar looked at Ivan. "Did you build the new one to the exact specifications as the old one."

"We still have the other plan," Ivan said. "Everything is exactly the same."

Bolivar shrugged. "Well, apparently it will work well if we need to use it. I would say what the Hidden Empire did was a really good test."

Leon looked at Ivan and motioned for him to continue his explanation of the status of the new hi-tech fortress.

"The walls are up. The roof is on. And all four floors are in place." Ivan glanced at his watch. "We're installing the AI capability you requested on the first floor as we speak."

"Are we still going to have two different floors with the AI capability?" Leon asked.

Ivan looked at Chen.

"Yes, I agree it's best to back everything up," Chen said. "We'll have the same AI setup on the fourth floor. That will be ready in a few more days. Since we used the same plans as were used for the first hi-tech fortress, everything will be the same, just better capability wise."

Leon nodded and didn't comment.

Ivan looked directly at Leon. "By tomorrow morning, the first floor will have all the furniture needed to utilize the AI capability." He kept a steady gaze on Leon. "By this time next week, the building will be completely furnished."

"But I'll be able to start using all of our AI capability tomorrow morning?" Leon asked.

Ivan nodded.

Chapter 66

Early the next morning in the main conference room at the FBI Headquarters in Jacksonville, Ben, Ralph, Matt, and Todd sat with Wendell, Dan, Vince, Justin, Paul, and Gary around the large table.

Wendell looked at the text on his phone. "Craig just notified me that NSA intercepted a phone conversation between a party in Moscow and a party here in Jacksonville. A suitcase nuke is coming into JAXPORT tonight."

Gary looked at Ben. "Are you prepared to handle it?"

"Yes," Ben said. "We have the handheld devices with the updated software, and we'll use the same team we've used before."

When Ben finished giving the details, his phone rang. He answered and listened for over a minute before he disconnected.

"As if things weren't complicated enough," Ben said, "our FBI network has just been hacked. Some of our experts are currently assessing the damage."

"Is there any way to tell if this cyber-attack was from a

quantum computer?" Matt asked.

"The caller told me that given a little time, they think they'll be able to verify the exact type of computer."

Justin leaned forward. "Just think what these bastards will be able to do if we don't stop them real soon."

Todd looked at Dan. "How will we know if this attack was initiated by a quantum computer?"

"Our team of experts in Washington indicated there have been some attacks in the last week where the polymorphic spyware could change its digital signature to millions of different combinations to evade identification by conventional anti-virus detection software," Dan said. "Our analytical team is using a quantum computer, so our detection capability isn't conventional. Very soon we'll be able to identify the exact type of computer used in the hacking."

Todd glanced at Ben. "Did the caller mention how the FBI discovered the attack?"

"He told me our quantum computer is connected to our defensive system on our network," Ben said. "It detected the hacking attempt and stymied the attack. Our assessment team checked for damage and found none. We're okay for now."

Todd nodded.

Ben glanced at his watch. "We need to move back to our top priority and plan our strategy for tonight." He looked at Matt and Ralph. "We'll do the recon like we did before, and I'll have a full SWAT team standing by to take over when necessary."

"Like last time, I'll set a command center up here," Ben said. "Dan, like before, will run the command center operation." Ben's gaze took in Justin and Vince. "If you're willing, I'd like the two of you to assist Dan again. Your knowledge on a lot of local issues will be a big help."

Justin and Vince both nodded.

Chapter 67

In Vienna, Austria, Zeus leaned toward Odin and Jupiter at the round conference table in a secluded room at their secret location.

Jupiter leaned forward and stared at Odin and Zeus. "Should we let Belzex continue with their plans to ship the suitcase nuke into JAXPORT?"

Zeus tightened his lips. "I think we should let them go ahead for now. We're going to destroy them later anyway." He paused. "It won't hurt anything for them to create a lot of disruption for law enforcement right now. That's good for us."

Odin grunted. "Belzex won't get the nuke through the port this time. They don't know it, but they no longer have the capability to block the signals on the nuclear detection equipment the FBI will be using at the docks."

"So, we don't need to worry about them detonating it in Jacksonville?" Jupiter asked.

"Not this time," Odin said. "For right now, NSA has leapfrogged ahead of them on AI capability. My hi-tech team is keeping me fully up to date on all that."

Jupiter narrowed his eyes and turned fully toward Odin "Hasn't Belzex just moved into their new hi-tech fortress in Germany?"

"Yes, and they've made a lot of improvements in their AI capability, but not as much as NSA has, and certainly not as much as we have," Odin said. "At the moment, we have by far the best AI. That's the reason I know these things."

There was a brief silence.

"Well, that brings us to the point of making an interesting decision," Zeus said.

Jupiter and Odin waited for Zeus to continue.

Zeus narrowed his eyes. "Will we allow Belzex to continue to exist, or will we give them a little more time?"

"Since they're now ready to move into their hi-tech fortress with all their advanced tools, they're probably going to make rapid progress in improving their AI capability," Jupiter said. "I say let's go ahead and destroy them now."

Odin looked at Zeus. "Well, what do you think?"

"It might be best to go ahead and eliminate them now," Zeus said. "I was just mulling over all the considerations involved." He looked back at Odin.

"I think it might be best too." Odin paused for a few seconds. "And I agree with Jupiter; they'll probably start making rapid progress in their AI capability."

"Okay," Jupiter said. "With our superior surveillance capability, we heard Leon tell how he survived last time. We'll use the same plan again and this time we'll make sure everyone is accounted for in the fortress just before we set off the explosives."

Zeus nodded. "Making sure everyone is accounted for in the fortress at the instant of the explosion is the best way of ensuring no one is in the bomb shelter."

Chapter 68

Later that afternoon in Germany, Leon stood just inside the doorway in the hi-tech control room on the first floor inside their new hi-tech fortress.

Belzex's other five members stood with him.

The control room was fifty feet long and fifty feet wide.

On the far wall, a computer screen extended from floor to ceiling, which was thirty feet high.

Four conference tables, each with chairs for eight people, were positioned about ten feet inside the doorway.

Ten tall chairs (the seats raised about four feet from floor level) spread horizontally across the room between the conference tables and the large screen.

Each tall chair had a small, attached desktop in front of it.

The tall chairs faced the large screen, which was twenty-five feet away from them.

The seat of each of the tall chairs was padded and each tall chair had two padded arms along with the small adjustable desktop. A sophisticated-looking black-metal control device was on each desktop.

Leon waved everyone toward the tall chairs.

Momentarily the group settled into six of them.

After glancing at Chen, Leon said, "You can get us started by explaining the details on how to operate all this."

Chen pressed a button on his control device, and the screen was immediately filled with images and written information. He started explaining the details.

Everyone else listened carefully and asked a lot of questions.

After two hours of discussion and practice, each of the men could expertly operate the controls in front of him.

"Okay, let's get to the business at hand," Leon said. He placed the red dot from his laser pointer on the image showing the site of the Tuxtun headquarters building in Jacksonville.

Everyone stared at the image the red dot was on.

"All of our satellites are in place all over the world and all of our other spying equipment is fully functional." Leon laughed. "We spent over a trillion dollars on this setup, and it looks like it's going to be worth it."

Everyone nodded.

"We, of course, can't know for sure, but we think the capability we have here rivals that of the Hidden Empire," Chen said.

"We have over ten thousand programs running at the same time on two quantum computers. Each program searches for specific things we want to keep track of and displays the images and information instantly," Ivan said. "We can also run specific searches at any time."

Chen nodded. "We can run an unlimited number of specific searches at the same time the AI is doing the searches we've programmed. We have no limitations."

Leon gestured toward the image he had focused his laser pointer on. "How will I disable the security sensors at JAXPORT when I need to?"

Chen explained.

"And how will I disable the security sensors at Tuxtun?" Leon asked.

Chen recommended they wait until they needed to get the nuke in place at Tuxtun before they did that.

Leon nodded.

There was a brief silence.

"Okay," Leon said. "Let's go back to when and how we'll disable the sensors they're going to use tonight at JAXPORT to detect the suitcase nuke?"

Chen explained.

"The suitcase nuke will be the ultra-powerful type with highly enriched uranium, right?" Leon asked.

"Right," Chen said.

Leon waited a few seconds and then said, "Let's review our planning. We can't allow any glitches in anything." He looked at Jake. "Have you heard anything more from The Hidden Empire?"

Jake shook his head. "Apparently none of our activities cause them any significant concern."

Leon nodded. "They obviously like the disruption we're going to be causing."

"Yeah, they've told us that several times," Jake said. "The disruption helps them keep their activities more invisible to the rest of the world."

Ivan leaned toward Jake. "And you told us they made it clear they'll get back in touch with us if they disagree with anything we're doing or planning to do, right?"

"Yes, they've made that very clear," Jake said. "They have all the means necessary to keep up with everything we're doing."

Leon gestured toward the multiple displays on the large screen. "We'll soon have the means to check on what the Hidden Empire is doing."

"We still have a lot of fine-tuning to do before we get to that point," Chen said, "but we're moving fast. The programming for both quantum computers is working to perfection. Our AI engineers and analysts have done a great job. We can instantly collect mountains of information about anything that concerns us. We'll constantly stay well informed as we go forward."

There was a brief silence.

"The main point is that we'll constantly improve fast," Chen said.

"Anything specific you're doing?" Leon asked.

"We're using a system where all the individuals on our team of analysts work in shifts to constantly monitor the information and immediately report on anything they think we should be notified about," Chen said. He gave a lot of details for several more minutes.

"We're in a great position to make fast progress," Jorge said.

Everyone gave a thumbs-up.

Chapter 69

At JAXPORT that evening, Matt focused his binoculars on a large ship that was in the process of docking at a long pier in front of them.

Equipment for unloading large crates was already in position alongside the ship.

Ralph crouched in the shadows ten yards left of Matt. Ben was in the shadows about ten yards to Matt's right.

Each member of the three-man team carried only a pistol for protection . Each man carried a Glock19 with a 15-round magazine of 9 mm bullets.

They planned to get in and out unnoticed. If they detected a nuke, a fully equipped SWAT team was waiting in an area nearby.

While staying in the shadows, Matt, Ralph, and Ben carefully watched the cargo now being unloaded, mostly large containers.

"They're fully in process. No inspectors," Matt whispered into his tactical microphone.

Yeah, as it was before. A lot of people have probably again

been paid off big," Ralph said. He paused. "There's now a hell of a lot of metal crates on the dock."

"If one of them contains a nuke, we need to find it." Ben reached into a pocket. "It's time to activate our detectors."

After removing a device from his pocket and hitting a button, Matt took a new position, still in the shadows, about twenty yards closer to the crates. No hits on his detection device yet.

The expected noise came from the activity along the pier. Everything else was quiet.

The three men stayed hidden and continued to carefully scan the area, visually and with their devices.

The crates on the ship continued to be unloaded.

The three men maintained constant communications with the command center back at the FBI Headquarters Building as they continued sweeping their devices back and forth across the dock area.

Momentarily, all three detection devices pinged.

"They probably had the same shielding as before, or maybe even better, but our detection devices lived up to the hype, so to speak," Ben said. "Let's find the specific crate."

Each man moved out of the shadows, kept a sharp lookout as he scanned the area, and followed the signal he was getting on his device.

In less than a minute, the three men surrounded the crate that obviously contained the nuke.

Ben punched a digit on his phone to alert the SWAT team.

The SWAT team leader answered the call.

"Okay," Ben said. "We have the nuke and we're ready for you guys to come take over."

"Roger," the SWAT team leader replied.

In another five minutes, Ben, Matt, and Ralph were heading back to FBI headquarters.

Chapter 70

The next morning at the FBI headquarters in Jacksonville, Ben and Kyle met in the main conference room with Ralph, Matt, Judy, Todd, Steve, Justin, and Vince.

"In a few minutes, Wendell, Kevin, Butch, and Dan will be joining us for the meeting," Ben said. "They took a CIA jet from Washington." He glanced at his watch. "They should be arriving soon." He started to say more just as the group from Washington arrived.

After everyone was settled, Ben asked, "Does anyone have any new intelligence information?'

"Quite a bit," Dan said. "That's the primary reason why we're here." He looked at Matt and Ralph. "We think we can help you with your security. Leon Schulz is now in his hi-tech fortress, and his AI is fully functional."

"Do you know if he's still trying to eliminate us?" Ralph asked.

"Yes he is," Dan said. "And he now has a lot more capability. NSA informed us that his AI software has all the latest updates."

"Is his AI capability now better than NSA's AI capability?" Matt asked.

Dan shrugged. "That's hard to know. NSA assures me they're trying to determine that. They're also working hard to improve our AI capability." He gestured to Wendell.

Wendell stared at Matt for a few seconds and then said, "I guess you're wondering if they're back to being able to override the detection capability in your security systems."

Matt nodded.

"I can tell you that Leon is now situated to be able to do that again," Wendell said. "I'm sure he's using his quantum computers to probe any weaknesses in your defenses."

"At least we got the nuke last night," Ben said. "We don't need to worry about dealing with that at the moment, although I'm sure he'll try to get another one shipped in."

"I'm guessing Leon is going to unleash some more assassins." Wendell gestured toward Kevin and Butch. "That's why these guys came back down here with us. They should be able to help in many situations."

Kevin and Bruce both nodded and didn't comment.

Steve mentioned some security issues he was concerned about and the discussion continued for another two hours with everyone giving their opinions about the best defensive strategy to adopt.

Chapter 71

That afternoon in Germany, Leon sat in the middle chair of the ten tall chairs that stretched horizontally across the hi-tech control room inside their new hi-tech fortress.

Belzex's other five members occupied some of the chairs beside him.

Leon used his control device to study various details on the large screen in front of him for several minutes.

The other Belzex members did the same.

In a few more minutes, Leon glanced at Ivan. "I think our top priority now should be to eliminate Baker and Gibson, and, this time, I want to use some of your assassins."

"I'm all set," Ivan said. "My assassins have all been briefed and they're ready to go." He paused. "Do you have any specific instructions?"

Leon narrowed his eyes. "Not right this moment. I thought we could all discuss it for a few minutes and come up with a super plan."

"My assassins are all well trained, but two of them stand out from all the others," Ivan said. "I think I'll use them for this next mission."

"Which two assassins are you choosing?" Leon asked.

Ivan leaned forward. "Stepan Petrov and Egor Ivanov are the best I have."

"How do you feel about keeping them?" Leon asked.

Ivan looked confused. He hesitated for a few moments and then looked back at Leon. "Why are you asking a question like that?"

"We haven't formed our plan yet," Leon said. "If they needed to be sacrificed, I was wondering how you felt about that."

Ivan narrowed his eyes and didn't comment.

Leon waited about thirty seconds and then leaned toward Ivan and asked, "Well?"

"They'll be valuable to have around for a long time as we go forward with our plans," Ivan said. "I certainly want to keep them around." He looked directly at Ivan. "I don't want to lose them."

Leon gave a short laugh. "Well, I think that's clear enough." He scanned the faces around him. "We'll keep that in mind as we're doing our planning."

Everyone turned slightly in the direction of Leon and nodded.

"The Hidden Empire doesn't seem to have any problem with our knocking off Baker and Gibson," Leon said. "So, the door is wide open for us to form a good plan and execute it. We're being given a great opportunity, and we don't want to miss out on it."

"And we won't miss out on it," Ivan said. "Like I mentioned, my assassins are my best ones, and, if we develop a decent plan, I have full confidence in them to get the job done correctly."

Leon grunted. "Okay, you've made your point." He paused. "Let's just hope they perform as we'll as we're expecting."

"Don't worry about that," Ivan said. "They'll succeed." He

paused and looked directly at Leon. "Remember, they're my two most skilled assassins, and I have a lot of good ones."

Leon didn't comment immediately.

"We're going to need to kill a lot more people as we go forward with our plans to conquer the world," Ivan said. "Egor and Stepan will be very valuable to still have around."

Leon gave a quick nod and glanced at his watch. "Okay let's get started on putting a plan together." He punched a few buttons on his control device, and a picture of the Tuxtun Headquarters Building occupied a section of the large screen in front of them.

After Leon punched a few more buttons, a floorplan appeared next to the building showing the first floor where Ralph Gibson's office was located.

"That's a live picture of the building, by the way," Leon said. "So, you might see some activity."

No one commented, although everyone was looking intently at the building.

Leon moved the red dot from his pointing device to the location of Ralph's office on the floorplan and then looked at Chen. "We'll know if they're in the office, right?"

Chen nodded. "We have all our capability functioning now. We can track them anywhere they go. Also, we'll constantly know who's in the office, and we'll be invisible to their security devices."

"Good," Leon said. "Now let's put a plan together, a plan that we're sure will work." He looked at Chen. "Is our AI capability still ahead of what NSA has?"

"I guarantee we're ahead of them again, and my team is working around the clock making sure we stay ahead of them going forward."

The discussion lasted for three more hours.

Chapter 72

Later that afternoon in the Jacksonville FBI office, Ben, Kyle, and Dan met in the main conference room with Ralph, Matt, Judy, Todd, Steve, Justin, and Vince.

Wendell, Kevin, and Butch from the CIA were also there.

Ben leaned toward Ralph and Matt. “I have an update from NSA. Two Belzex assassins are on their way here. NSA was able to intercept and decrypt several phone conversations. We picked up a lot of information.”

“Did you get any information about the assassins?” Ralph asked.

“Quite a bit,” Ben said. “They’re both Russian, and their names are Stepan Petrov and Egor Ivanov.” He gestured toward Kevin and Butch.

“The CIA has their profiles and Butch and I have studied them,” Kevin said. “We even have their photographs.”

“Is this the same Petrov who was in the group with Leon when we had the meeting with Belzex about buying the land?” Matt asked.

Ben nodded. “Same one.”

Kevin displayed the photograph of each assassin as he said, “We know all about them and their skill levels equal ours. We’ll need to be careful.”

Ralph looked at Todd. “Are you keeping our security system up to date?”

“Of course,” Todd said. “Kyle and I just installed the latest updates from NSA before we began this meeting."

There was a brief silence.

“Do we know how good our AI technology is right now and how it might compare to that of Belzex and the Hidden Empire?” Matt asked.

Wendell glanced at Dan.

Dan gestured for Wendell to go ahead and address the question.

“Most any organization’s AI capability changes constantly,” Wendell said. He scanned the faces around him. “That makes it almost impossible to get an exact measurement at any point in time. Right now, we think we’re again ahead of Belzex.”

“So, all of our sensors should be working on our security system?” Matt asked.

“We think so, and we’re using every way we’re aware of to verify that,” Wendell said.

“Is NSA continuously working on getting an exact measurement of Belzex’s capability?” Ralph asked.

Wendell gave a slight laugh. “I can assure you of that. Whenever they get ahead of us, we want to know it.” He paused. “All indications from our extensive testing indicate NSA is ahead of each organization we’re keeping track of except, of course, for the Hidden Empire.”

“Which organizations are you keeping track of right now?” Ralph asked.

“We’re keeping track of many foreign countries including Russia and China. We’re also keeping track of various organizations in the United States including the FBI, the CIA, NSA, Tuxtun, and Belzex,” Wendell said. He paused. “At least we try to keep track of some things the Hidden Empire is doing."

Dan nodded. "Some days we think we're making some progress."

"So, right now, do you consider the Hidden Empire the enemy?" Judy asked.

"We don't really know how to classify them. We know they can do great harm to us if they choose to." Wendell shrugged. "We to try to keep up with their AI capability." He glanced at Dan.

Dan nodded. "Nothing to add, we're doing our best to try to find out anything we can about them."

"If we can ever find a way to trace the cashflow in the Specter Matrix, we'll be able to understand a lot of things," Wendell said. He glanced at Judy and then went on to give more details.

When Wendell finished, he glanced at Judy again. "Do you have anything to add?"

Judy shook her head. "You covered it well."

The discussion continued for a few more minutes before Ben suggested concentrating on some more imminent things, such as tracking the location and activity of the two assassins that were on their way to Jacksonville.

Chapter 73

In Vienna, Austria, Zeus leaned toward Odin and Jupiter at the round conference table in a secluded room at their secret location.

"So, we agree to eliminate Leon along with all the rest of Belzex," Zeus said.

Odin and Jupiter nodded.

"I've instructed Thor to make all the preparations," Zeus said. "He'll let me know as soon as everything is ready."

"Will he use the same team we had before to get in, set the explosives and get out?" Odin asked.

"Definitely," Zeus said. "They did it perfectly last time and they have the benefit of having that experience." He paused. "The new fortress was built using the same blueprint as the previous one. Thor doesn't expect to have any problems with any aspect of the mission." He looked at Odin. "You're responsible for our team that maintains our AI. Do you think we're still ahead of everyone else?"

Odin nodded. "I just checked with the team leader an hour

ago. He told me all the tests he continuously runs show that we're still ahead."

"All that stuff changes often. Is he keeping in close touch with Thor?" Jupiter asked.

"They contact each other every hour," Odin said. "And they plan to talk immediately before the team goes in to plant the explosives."

Zeus nodded. "And I'm sure Thor will make sure no one is in the bomb shelter when we're ready to detonate the explosives."

"That's a key part of his checklist," Odin said. "I don't think you need to worry about that this time." He gave a thumbs-up.

Zeus and Jupiter also gave a thumbs-up.

Chapter 74

At the Jacksonville FBI office, Ben continued his meeting in the main conference room.

Ben, Kyle, Dan, Ralph, Matt, Judy, Todd, Steve, Justin, Vince, Wendell, Kevin, and Butch were still sitting at the large table.

Ralph scanned the faces around him and then asked, “Does anyone have any updates on the status of the two assassins?”

Wendell grunted. “NSA tracked them to Jacksonville. They’re having a little trouble locating them, but they’re working on the problem and will keep us posted.”

“Trouble with some of their sensors being blocked?” Matt asked.

“Yes,” Wendell said. “Remember any organization’s AI capability changes constantly. At the moment, NSA can’t get any readings from their surveillance satellites. The satellites are their main tool for picking up exact locations of individuals” He paused for several moments. “They think Belzex has made some advanced updates to their overall AI system, but NSA is confident they can make the necessary improvements in our overall system within the hour."

Matt nodded slowly and didn't comment.

Kyle leaned toward Wendell. "Are you talking directly to someone at NSA."

"Yes," Wendell said. "I'm staying in close contact with the person who's overseeing the work. He calls me every five minutes to give me some updates."

"Maybe you four technical guys should go to another room and do conference calls with a couple of the people at NSA," Ben said. "We would get a better idea of the exact problems they're working on, and maybe you would have some helpful suggestions for them."

"That's not a bad idea," Kyle said. He stood to leave.

Wendell, Dan, and Todd also stood.

Chapter 75

At the Belzex hi-tech fortress, Ivan answered the phone, and after a short conversation, he looked at Leon. "That was Stepan. He and Igor are in Jacksonville and about ready to start their mission."

Leon looked at Chen and Ivan. "Are we all set? Will signals from our AI prevent our assassins from being detected?"

Chen laughed. "Yes, we're all set. Not only are we blocking the signals of the Tuxtun security system, we're also blocking the signals of NSA's surveillance satellites. We're in complete control. Baker and Gibson will never know what hit them."

Leon also laughed. "It will be great to have them out of our way." He paused and looked at Ivan. "How soon will your assassins be ready?"

"They're getting ready to leave the hotel where they were staying," Ivan said. He glanced at his watch. "They're walking since the hotel is less than a mile from the Tuxtun headquarters building. They should be there within the hour. They're going to call me again when they're on their way."

Leon nodded as he also glanced at his watch.

*****

At Tuxtun, Wendell punched a button on his phone where he had the speaker function activated.

When he heard the expected person answer at NSA, he asked, "Are you making any progress?"

"A little," the person said slowly. He paused for a moment and then gave a few details.

"And you're sure the signals are being blocked?" Kyle asked.

"We're sure they're being blocked by Belzex," the person said. "We've been able to trace the problem back to them. They've apparently improved their AI capability quite a bit."

Chapter 76

In Vienna, Austria, Zeus, Odin, and Jupiter continued their discussion at the round conference table in a secluded room at their secret location.

"It looks like Belzex is going ahead with their plans to in spite of our warning," Jupiter said.

Odin grunted. "That's not a big surprise to me. I don't think Leon can stop, even if he wanted to. He obviously has an irresistible drive to conquer the world and we're in his way."

"But first, he needs to eliminate an immediate obstacle to carrying out his obsession," Zeus said.

Odin and Jupiter nodded and didn't comment.

There was a brief silence.

Zeus leaned toward Odin. "Does our technical crew have any additional news?"

"Tuxtun's financial expert along with several technology specialists have teamed together to try to track the cashflow in the Specter Matrix," Odin said.

"Do they know if Tuxtun's making any progress on that?" Jupiter asked.

"The leader of our team tells me we have no worries in that regard," Odin said. "He assures me no one will ever get close to being able to track that cashflow no matter how good their AI capability is."

After Odin explained a few details, Zeus and Jupiter nodded.

As soon as he finished explaining about the Specter Matrix, Odin reached into one of his pockets and retrieved his vibrating phone. He checked the caller ID and answered.

In a few moments, Odin disconnected and looked at Zeus and Jupiter.

Odin narrowed his eyes. "Thor informed me the explosives have been set in place at the hi-tech fortress, and our team is out of the area."

"Okay," Zeus said. "I'm sure Leon is never going to give up trying to conquer the world and, with the resources they have, they'll continue to get more powerful, maybe even as powerful we are."

Jupiter nodded. "We're basically in control of the world, but we're a lot different than Leon wants to be. We work behind the scenes and influence things. We don't want to be to be out in the open and have one of us recognized as the emperor of the world by everyone on the planet."

"Right," Zeus said. He looked at Odin. "Tell your team to go ahead and detonate the explosives when they're ready."

Chapter 77

At the Belzex hi-tech fortress, Ivan removed his vibrating phone from one of his pockets and answered. After a short conversation, he disconnected and looked at Leon.

"That was Stepan," Ivan said. "He and Egor have left the hotel and are walking toward Tuxtun. They'll call back when they get there and make sure everything is still functioning properly with our AI capability before they go inside."

Leon nodded and looked at Chen. "We're still all set and the signals from our AI will prevent the security at Tuxtun from detecting our assassins, right?"

"I'm keeping in close touch with my person supervising the technology development, and he's reported no problems," Chen said. "I'll give him another call right now."

Leon nodded and didn't comment.

Chen punched a digit on his phone and quickly engaged in a conversation.

In less than a minute, Chen disconnected and looked at Leon.

"He told me everything is still great," Chen said. "Instead

of taking time to give me the details, he suggested I check them myself. He told me he had just updated our database that shows the status of everything. Give me a moment and I'll check that right now."

Leon nodded and gestured impatiently for him to keep going.

Chen studied a device he carried with him for a few moments and then looked back at Leon.

Everyone focused their gazes on Chen.

"We're still in complete control," Chen said. "Baker and Gibson will never know what hit them."

Leon laughed. "It will be great to have them out of our way. You can't beat the feeling. We'll finally get the land we need.................."

Chapter 78

Outside of the Tustin headquarters building, Stepan and Egor slowly walked toward the entrance. They wanted to complete their conversation and make a phone call before they entered.

Both assassins were the ultimate professionals. They wanted to check some more details and not leave anything to chance.

"Since I met them in the meeting we had at Tuxtun, I can recognize baker and Gibson instantly from any angle," Stepan said.

Egor nodded. "I've studied their photographs. We'll need to react instantly when we recognize them. They're damned good."

"We'll have an advantage. We'll have our pistols in our hands when we walk through the doorway to Gibson's office." Stepan glanced at Egor. "Since they won't know we're there, we'll have time to pause and draw just before we charge into the office. "

"Good plan, we would attract too much attention walking down the hall with drawn pistols." Egor nodded. "We'll make sure there are no people in the hallway when we get to the doorway into Gibson's office and are ready to charge in."

Stepan gave a thumbs-up. "We each have a 15-round magazine of 9x19mm Parabellum cartridges. We 're as ready as we can be."

Egor narrowed his eyes. "We should only need one shot each, but it good to have a lot more bullets, just in case."

Stepan gave a short laugh. "That never hurts."

"Any questions about the strategy we'll use when we walk through the front doors?" Stepan asked.

"I think we're all set. We won't stay close together, but we'll go straight down the hall to Gibson's office."

Stepan nodded. "You lead the way, and I'll trail about five yards behind you. You can wait a few seconds outside the doorway for me to catch up before you go in."

Egor gave a thumbs-up as both men walked over to a bench outside the entrance.

Stepan reached for his secure satellite phone. "I'll call Ivan and make sure everything is still on go."

*****

In a small conference room at Tuxtun, Wendell punched a button on his phone where he had the speaker function activated.

When he heard the expected person answer at the NSA, he asked, "Can you give me a status update?"

"It looks like we just got everything back. Give me a minute to check everything.

Wendell gave a thumbs-up immediately.

In less than a minute, the person at NSA was back on the phone.

"Not only are our satellite sensors working again, but all of your security sensors at Tuxtun are working."

"I didn't know our security sensors weren't functioning," Wendell said.

"They weren't for a while," the NSA person said. "I guess Belzex found a way to do that without any alarms going off." He paused. "There's something you should know. There are two assassins sitting on a bench right outside the entrance to your building. Their names are Stepan Petrov and Egor Ivanov."

"Thanks," Wendell said. "Are you sure all of our security systems are working?"

After a short pause, the NSA person said, "According to all of our indicators, everything is working perfectly."

"Thanks again," Wendell said as he stood and gestured to everyone. "We're going back to the boardroom."

*****

The four technology experts returned to the boardroom at Tuxtun and gave a quick update.

"Well, it's good we're fully functional again," Ben said. He looked at Todd "Have you checked to see if the two assassins are still out there on the bench?"

Todd scanned the faces around him. "They're still out there. I just checked again. Each one is using his phone at the moment."

Ben started to comment when his phone vibrated. He checked the caller ID and answered.

*****

Stephan looked at Egor. "I'm wondering if you got the shocking news I just received?"

"Probably," Egor said. He narrowed his eyes. "My boss in Moscow told me that Belzex has been completely wiped off the map."

Stepan nodded. "Ivan doesn't answer his phone and that's why: He's not there anymore."

There was a brief silence.

"Everyone is gone," Stepan said.

Egor regarded Stepan carefully. "Unless you know something I don't know, there's no need for us to continue this mission."

"Agreed," Stepan said. "I don't personally like Baker or Gibson, but I don't want to go to the trouble to try to kill them just because I don't like them."

Egor nodded. "Same here. Besides, their security sensors are probably not being blocked now. We'll be detected."

"We'll call our bosses again and have them make arrangements to get us back to Moscow," Stepan said.

Both men stood and walked back toward the hotel.

*****

In the boardroom at Tuxtun, Ben rubbed his chin and scanned the faces around the table.

"We've received a lot of new information from NSA that we should discuss," Ben said. "Most of it involves advanced technology." He gestured toward Wendell. "I'll let our technology group lead the first part of the discussion and then we'll discuss some other aspects of what just happened, including some financial aspects."

Wendell studied his laptop computer for a moment and then said, "The experts at NSA have concluded it was the Hidden Empire who wiped out Belzex."

Matt noted that all four technology experts were staring at their laptops and appeared to be concentrating on the new data they had just received. He also looked at his laptop and studied the new data.

A discussion on the technology involved in the situation lasted for over an hour.

"Okay," Ben said. "I don't think we can wrap up this discussion yet without getting into some broader aspects of the current status of things." He glanced at Ralph and Matt. "One of those aspects concerns the possibility of pending danger to Tuxtun."

Ralph nodded. "We know a lot about the Hidden Empire now and we've seen three times how they eliminate threats to their existence. We don't want to get on their wrong side." He looked back at Ben. "Any suggestions?"

"The thing that initially comes to mind is for all of us to quit trying to trace the cashflow through the Specter Matrix," Ben said. "NSA is doing it." He looked at Judy. "And you're doing it along with all the help you're getting." He made a sweeping motion with his right hand. "And we're all involved in some way."

"If the Specter Matrix stops working perfectly, that takes away the lifeblood of the Hidden Empire's existence," Matt said. "And we all know the Hidden Empire is not going to let that happen."

"As the saying goes: we're between a rock and a hard place," Ben said. "We all need to keep doing what we've been doing in that regard." He paused. "NSA must continue to analyze global cashflow in order to identify any illegal activity that needs attention. All law enforcement has some involvement in that."

Everybody nodded.

Ben looked at Ralph. "I know you've had reasons for Judy and Todd to be involved in that too, but now that Belzex is gone you might want to rethink that for your safety."

"We will," Ralph said. He looked at Judy and Todd.

"We'll still want to be knowledgeable about the global cashflow methods used by drug traffickers because, in a way, we're in competition with them."

"Drug traffickers buy land to grow poppies to supply their needs for legal and illegal products," Matt said. "Sometimes we run into some complications that require our attention in several ways, business wise and finance wise."

"And security wise," Steve added.

There was a brief silence.

Ben looked at Judy. "I'm just curious, did you ever make any real progress in trying to trace the cashflow through the Specter Matrix?"

Judy gave a slight shrug. "The short answer is no." She paused. "The longer answer is that we thought we were making good progress many times, but we always came up empty." She gestured toward Todd. "He and the other technology guys provided a lot of help, and I want to thank them for that."

Todd, Kyle, Wendell, and Dan each gave a thumbs-up.

"We learned a lot, and we'll continue to learn, but it seems the Hidden Empire always stayed one step ahead of us on AI capability," Todd said.

Kyle, Wendell, and Dan all nodded.

The discussion lasted for another thirty minutes before Ben said, "Well, I guess that's enough discussion for now. We'll all stay in contact on things of significance as we go forward."

Chapter 79

The next day in Vienna, Austria, Zeus, Odin, and Jupiter continued their discussion at the round conference table in a secluded room at their secret location.

"Well, Leon never even slowed down on his plan to conquer the world," Zeus said. "He didn't learn his lesson when he got a second chance and he paid the price."

Odin and Jupiter both nodded.

"I guess the genes he inherited made the urge he had to conquer the world irresistible," Jupiter said.

Zeus shook his head slowly. "Apparently that was the case for both him and Hugo."

There was a brief silence.

"Now is a good time to make sure we're okay on all fronts as we go forward," Zeus said. He looked at Odin. "You're constantly checking with our advanced technology leader and our finance leader. Do you have any comments about our status?"

"Our leaders in those two areas are still working closely together and producing great results," Odin said. "We're concentrating on the Specter Matrix and trying to stay ahead of everyone else."

Zeus gave a short laugh. "And it looks like you're doing that."

Odin nodded. "Since Belzex is gone, NSA is our biggest concern, but, so far, our experts are staying ahead of their experts."

"You're the most qualified among the three of us in those areas," Zeus said. "Don't get into too much technical detail but tell us a little bit about what our experts are doing to keep us in the lead."

Odin leaned forward. "We've talked before about many of the things we're doing." He paused for a moment. "I guess it would be good to mention some of the newest things we're using to stay ahead of the crowd, so to speak."

After spending the next twenty minutes explaining some new tools and techniques, Odin asked if there were any questions.

"So, PKI stands for Public Key Infrastructure and is the fundamental cybersecurity framework for securing web traffic?" Jupiter asked.

Odin nodded. "It relies on asymmetric cryptography." He gave some more explanation.

Zeus glanced at his notes and then looked at Odin. "And QKD stands for Quantum Key Distribution?"

"Yes, it uses quantum mechanics to alter quantum states to detect an intrusion immediately," Odin said.

Zeus shook his head slowly and looked back at his notes. "PQC stands for Post-Quantum Cryptography and uses complex mathematical algorithms to detect intrusion?"

"That's correct," Odin said. "And it can be run on all computers which is beneficial since the paths the Specter Matrix takes varies a lot throughout the financial global network."

Jupiter looked up from his notes. "I have one final question." He checked his notes again. "ECC stands for Elliptic

Curve Cryptography and it's an encryption algorithm that uses algebraic geometry of elliptic curves to secure data?"

Odin gave a short laugh. "That's a mouthful and you got it correct."

There was a brief silence.

Odin scanned the faces around him. "Any more questions?"

Zeus and Jupiter remained silent.

"I figured it might be a good thing to introduce you to a little bit of the terminology so if you hear some of it from our advanced technology team, you might have a general idea of what they're talking about," Odin said.

Zeus shrugged. "I think we're going to leave communicating with them up to you. Jupiter and I don't have your expertise in the subject matter."

Jupiter gave a thumbs-up.

Odin smiled and gave a slight nod. "Well, the bottom line is that the activity in our Specter Matrix is secure in every way possible, and we'll keep it that way."

A brief silence ensued.

"The Hidden Empire has existed for a long time, and it will continue to exist for a long time," Zeus said. "Having a lot of power has always been our main goal and that will continue to be our main goal." He paused. "We started out with a lot of wealth and we've invested well. We've planned well too. We've created the Specter Matrix to help us make timely payments to all the right people and organizations as we move forward."

Jupiter nodded. "We have the wealth, the organization, and the tools that we need."

"And the main tool is the Specter Matrix," Odin said. "We have the means to keep it updated and immune to any detection by any other organization."

Each man raised his right fist high into the air.

www.ingramcontent.com/pod-product-compliance
Lightning Source LLC
LaVergne TN
LVHW052038200926
848481LV00005B/1715